The Ever-Rolling Stream

"He that publishes a book runs a very great hazard, since nothing can be more impossible than to compose one that may secure the approbation of every reader."

Miguel de Cervantes

The Ever-Rolling Stream

BERNARD ALDRICH

Illustrated by Manuela Smith

London
GEORGE ALLEN & UNWIN
Boston   Sydney

© Bernard Aldrich, 1984
This book is copyright under the Berne Convention. No reproduction without permission. All rights reserved.
George Allen & Unwin (Publishers) Ltd
40 Museum Street, London WC1A 1LU, UK

George Allen & Unwin (Publishers) Ltd,
Park Lane, Hemel Hempstead, Herts HP2 4TE, UK

Allen & Unwin Inc.,
9 Winchester Terrace, Winchester, Mass 01890, USA

George Allen & Unwin Australia Pty Ltd,
8 Napier Street, North Sydney, NSW 2060, Australia
First published in 1984

British Library Cataloguing in Publication Data

Aldrich, Bernard
　　The ever-rolling stream.
　1. Fishing—England—Test, River
　2. Fishery management—England—Test,
River　3. Broadlands (Romsey, Hampshire)
I. Title
799.1'1'0924　　　　SH606
ISBN 0-04-799019-8

Set in 11 on 12½ point Palatino by Computape (Pickering) Ltd,
and printed in Great Britain by
Butler & Tanner Ltd, Frome and London

*To my wife, Marie,
and daughters, Mary and Jill,
for their love, help and support*

Foreword by
H.R.H. The Prince of Wales

Ever since I was a young boy Bernard Aldrich has been an important feature of Broadlands. His cheerful personality, his sense of humour and his optimistic approach to fishing made him a wonderful companion on the river, as well as being someone from whom it was a great pleasure to learn about the natural features of The Test. I suppose the first time I really recall his presence in my childhood memories was when my great uncle, Lord Mountbatten, took my sister and me down to the river to see if we could find a salmon lying under a bridge. The resulting story is well described by the author in a later chapter of this book, but it was the first of many happy hours I have spent with Bernard trying to inveigle the occasional salmon out of the pools on the Broadlands stretch of The Test.

Such was the fun we used to have, and so fast did the time pass, that I was invariably late returning to the house. All was well if I actually returned with a contribution to the larder, otherwise I received a "rocket"!

I am so glad that Bernard has decided to write this book for it helps to show those of us who value the countryside and its wildlife just how much we owe to people like him. Their knowledge, experience and deep love of nature, coupled with the long hours of dedicated toil, help to preserve some of those features we appreciate most. Man has interfered with nature so thoroughly that it is now even more essential that such people as Bernard should conserve and manage the local environment and wildlife; should warn us, as he does, of the dangers to a river system like The Test and, above all, should be listened to as intelligent observers who often instinctively *feel* something far more accurately than any scientist can subsequently prove it.

This book is written with an obvious and deep-rooted love for the river, by whose banks the author has lived for twenty-five peaceful years. It is a happy book, written by a contented, kind-hearted and loyal man and after you have read it it will, I assure you, leave you with an equally happy and contented glow.

Charles.

CHAPTER ONE

I WAS BORN close to a river. In fact, about the same distance as I now live from the Test. But it was a totally different river from the Test in every way. It was the Thames at Woolwich in south-east London, and the year was 1928. All the London docks were thriving, with ships large and small continually passing up and down and depositing their effluent into the water as they went. As a result, the river was grossly polluted and held no fish life of any sort. When the tide had ebbed, large areas of foul mud were exposed and the stench was awful. I well remember that during the summer, my mother would have to keep all the windows and doors closed, as otherwise the smell would spread throughout the house and linger for days in curtains and soft furnishings.

There was a superstition among local mothers that if a child had whooping cough, a miracle cure would be to take the suffering infant on several trips across the river aboard the Woolwich Free Ferry, allowing the poor child to breathe in the "sea breezes". It was, I think, a miracle if the child survived this kill or cure treatment.

I was the third of four children, the only boy. My father was at sea with the General Post Office, employed as chief jointer aboard the telegraph ship *Monarch*, laying submarine cables. Although the ship's depot was in Woolwich, we did not see a great deal of father, but he was a lovely man and we were a happy little family. Tragically, our mother died when I was seven years old, but my elder sisters stepped into the breach to care for the two youngest of the family, so taking on the duties of full-time mother and part-time father. They did a magnificent job under very difficult circumstances, and I must have been quite a handful as a youngster.

In 1939, when the second world war began, my younger sister and I, along with the majority of our schoolmates, were taken to the railway station with our rather pathetic bundles of "personal effects". With name-labels pinned to our chests, we were packed onto a train and evacuated to the country – Goudhurst in Kent. The intention was that London children should be safely away from the city when the expected bombing started. The bombing did not start, for this was the period of the so-called phoney war, the lull before the storm.

My sister and I were not very happy being evacuees. It wasn't that we disliked country life or the country people. We were just desperately homesick. So eventually we were brought back to London – just in time for the Battle of Britain and the Blitz! As a boy I found it all very exciting and somewhat frightening but by now, being eleven years old, I was at a senior school, which I didn't like at all. I played truant with a friend and helped a milkman on his rounds, who gave us enough money to visit the cinema every afternoon. I think we saw *The Sea Hawk* about twelve times and never tired of it. All I ever dreamed about was going to sea. My father was rather against it, probably because he knew what a disrupting effect sea life had on a family. However, he did finally relent and at thirteen years old I joined the training ship *Indefatigable*. She was in fact a ship based on the River Mersey, but because of the bombing on Liverpool had been evacuated to a shore establishment at Clawdd Newydd in North Wales.

It was a hard life on the *Indefatigable*. We wore wooden-soled clogs with tough leather uppers, and short trousers all the year. The hill-tops in Wales can be very cold in winter, especially on small boys' knees, but in spite of it all, I loved it. I was fortunate enough to be chosen to go on a course at the Outward Bound Sea School at Aberdovey, the same one that Prince Philip had attended a few years before me. Part of the course was spent aboard a lovely old schooner, *Prince Louis*. She was a fine ship, but without any mechanical aids; everything was "handraulic", as the old skipper described her winches and working gear. Whilst learning to sail *Prince Louis* we trawled for fish to supplement the wartime rations at the school. Our arms were permanently swollen from the jelly-fish stings which had adhered to the trawl ropes and transferred to us as we hauled in. I was soaking wet most of the time, being "bowsprit boy", which meant that I hung out over the bow releasing the jibs. As the ship rose and fell, I was plunged waist-deep into the sea. It certainly did me no harm, although I didn't really appreciate it at the time, and I never showed any sign of getting a cold.

At last the big day came when I left the training ship and, at the tender age of fourteen, reported to the "Pool Office" which was the Merchant Navy's Labour Exchange, dressed in my new blue serge suit that had been supplied to replace my "Indefat" uniform. There I was issued with a discharge book in which was to be recorded each sea voyage, my special identity card, complete with photograph and fingerprints, and finally a small silver "M.N." lapel badge. These badges were given to merchant seamen to distinguish us as men doing our bit for King and Country, for we wore civilian clothing and were liable to harassment from the populace for being cowardly without such identification. I treasured my badge. I have it to this day and would not part with it.

I was ordered to report to the S.S *Empress of Scotland*, lying in Gladstone Dock. She was the ex-*Empress of Japan*, owned by the Canadian Pacific Line, a ship of 27,000 tons and holder of the coveted Blue Riband of the Pacific. Now she was painted all over grey and employed as a troopship.

Being at sea was everything I had dreamed of: exciting, frightening and very hard work! I enjoyed every moment. I met some very fine men and made a lot of good friends. During my sea-time I served on a great variety of vessels: troopers, tankers, tramps, Royal Mail and at the end of the war a super luxury liner, the *Dominion Monarch*. She was launched in 1939 and never had a chance to do the job she was built for, luxury cruising to Australia and New Zealand. Instead she was painted wartime grey and went straight into service as a troopship. At the time she was the world's most powerful motor ship and the largest refrigerated ship. It was whilst on leave from her that I met, or rather re-met, my wife Marie.

I had first met Marie as a cousin of my best friend. At the time he and I could not really be bothered with girls. Marie and he had an aunt living near Northolt Airport, which was a fighter base, and all we wanted was to be there and see the Spitfires take off to do battle. There was very little interest on our part in silly females. In retrospect, I am sure the "silly females" felt exactly the same about the stupid boys who only had eyes for aircraft. You must bear in mind that at the time we were only thirteen years old! But I think that because of necessity, we were much, much more mature in many ways than present-day thirteen-year-olds. Marie and I promised to write to one another, and we did for a while, but Marie lost the list of addresses she must write to when I was at sea, and that was that.

It was when I was returning from leave, having visited my married sister in Berkshire, that I was standing on Charing Cross Station waiting for my train home to Woolwich and a very lovely girl said to me, "Hello, Bernard. Remember me?" It was Marie, the cousin of my best friend. I took her home and asked her out again, and we were very soon "courting strong". I always say she picked me up on Charing Cross Station! It was the start of a very happy life together.

We planned to marry and emigrate to New Zealand, a country which I had visited many times and grown to like very much. I had also made a lot of friends there. Marie's aunt lived in Christchurch, New Zealand, so we would both have been well received. We discovered that it would be best to emigrate first and then marry, once we were established in our new country, so I resigned from the Merchant Navy and signed on a brand-new ship which I would take out to New Zealand for their government on a one-way trip contract. Marie had already arranged to go under the emigration scheme, which at the time had a three-year waiting list as there was only one ship operating. At last we both received our sailing dates, but a short while before I was due to leave I contracted an eye problem which prevented me leaving. Marie cancelled her passage, and we decided that as we would have to wait for another three years before we could sail for New Zealand, we would settle down in England.

We were married in 1950 at Dartford, Marie's home, and moved into the house that I had been born in. We lived with my father until he remarried, when we had the whole house to ourselves. We could not believe our good fortune in having a home of our own so early in our married life.

My problem was finding work. Anyone resigning from the Merchant Navy at that time was not allowed to re-join for at least two years. I was a very good sailor but that did not really stand me in good stead for job-hunting ashore, as I had no trade skills other than being a seaman. I tried painting and decorating but wasn't happy working indoors. Then I applied to and was accepted by the police force, operating in and around the Royal group of docks. Although it was a tough and rough area to be in, it was good to be in the open air again, and I quite enjoyed being a policeman.

With the Coronation of our Queen taking place, I was one of the officers from our division drafted into the City of London to line the royal route. I was on duty at the Mansion House, so had a grandstand view of the celebrations.

In the meantime my younger sister, Win, had married her husband, Len. They lived in Romsey and his parents would give an evening meal to a widower friend of theirs, one Walter Geary, who was head water bailiff for Lord and Lady Mountbatten at Broadlands. There was no television in Hampshire in those days, while we in London were the proud owners of a nine-inch TV set with a magnifier hanging over the screen, making it the equivalent of a modern twelve-inch set. The Romsey side of our family were invited up to London to watch the Coronation on our TV.

The contingent arrived, all seven of them including Walter Geary, crammed in one car, with bedding, food and luggage. The sleeping arrangements in our tiny house were cosy, to say the least. My brother-in-law, Len, Walter and I slept in the downstairs front room, Len and I on camp beds, Walter on the settee. We tucked up for the night and snuggled down to sleep. Len and I slept for about ten minutes before the room started to reverberate with the most awful thunderous snores emanating from Walter. Len and I just did not sleep a wink. In the morning we asked Walt how he had slept. "Not too well," he said, and refused to believe that his snores had kept his room-mates awake all night.

The Coronation viewing was a lot of fun, enjoyed by the whole family. It also began a friendship and working partnership between Walter and me which lasted for nearly twenty years.

I resigned from the police force as I could not understand the unswerving attitude of my superior officers when I wished to change duties with a colleague, enabling me to be off night-duty during the last days of my wife's pregnancy. Instead, I was lucky enough to find a position at the Royal Society of Medicine at 1, Wimpole Street. This was very unusual but interesting work. I was one of a team of those involved in showing films, slides and photographs at meetings to members of the Society. These films were of new or rare operation techniques, treatments, follow-ups, etc. Probably the one I remember best was Professor Ian Aird's operation separating Siamese twins with a common liver. They were little African babies known as the Boko twins. The operation was fascinating to watch, and the time, practice and skill involved were incredible. Sadly, one of the twins died but as far as I know, the survivor still lives, thanks to the dedication of a truly wonderful team.

By now we had two young daughters, Mary and Jill. We could not afford "proper" holidays so we took ourselves off to stay with my sister in Romsey. We all loved the country around Romsey and had many enjoyable forays into the New Forest. I particularly liked to visit Walter Geary on the river and to watch him fishing. One day I actually saw him kill a salmon. I think it was at that moment when I was hooked.

I was not a fisherman and had never even been interested in fishing, possibly because I had no opportunity to be near water that held fish, other than the sea. My one excursion into fishing had been while anchored in Mombasa harbour. The arc lights over the companionway onto the ship had attracted many hammerhead sharks. On the advice of our first officer we soaked a piece of cloth in blood from meat in the fridge, hooked it on to a sharpened grapnel hook and tossed it into the sea. It was taken at once by a sixteen-foot hammerhead. A few turns round a winch with the line and the fish was played, being finally finished off by the "Old Man" (the captain of the ship) with a .303 rifle. The teeth of the shark were removed and shared out amongst the "fishermen", and the meat was given to the natives. That had been my only experience of fishing of any kind.

The river Test was different from anything I had ever known and to watch Walter cast a fly, as only he could, was sheet poetry. I became entranced with the whole world of salmon and the river, the only water that I knew of then to hold salmon.

I spent more and more time on that holiday with Walter, until one day he suggested, no more than half-seriously, that I should work with him as his understudy. Walt was then aged sixty-five, and Lord and Lady Mountbatten had been rather worried that Walt might become ill or even die without passing all his wonderful knowledge of the river onto someone capable of using it for the future. I rushed home to Marie for her opinion. Bless her heart, she said she would back me in whatever I wanted to do. I felt then and do now, with the experiences we have gone through, that it is essential when decisions which could affect a marriage or life together are taken, it *must* be by mutual agreement and whatever the outcome, there should be *no* looking back. Rather, make the best of the way which you together have chosen.

Luckily for us, it was a spur-of-the moment decision. I can't help thinking that, had we had more time to consider the really giant step we were to take, we might have chosen to stay in London. We must have been among the first people to opt out of city life and, looking back, we weren't very well qualified to do so. Neither of us had ever lived in the country or knew anything about country folk and their ways. We had two very young children, and I was going to a job I had no experience in, not knowing whether I would like it or, more importantly, whether I would be liked by the fishermen and fit into their environment.

With the slight madness of youth, our minds were made up and I talked seriously to Walter, asking him if he would really consider me as his understudy. To my delight, he agreed at once and took me to the Estate Office to see the agent. He was not in the office, so I was dragged across to see him at home.

The agent at this time was Commander W.F.G.North, a very tall, impressive man, who invited me into his study. I was asked about previous experience, and would I be prepared to come to Broadlands at a later date for an interview. He then wrote my name and address on a tiny scrap of paper which he tossed onto his huge desk, which was covered in books and papers of every size and shape.

"That's that," I thought. Even if he wanted to, he would never ever find my piece of paper amongst all the debris on his desk. About a month went by and I really did think my name had been lost, when a letter arrived asking if I would be available for interview the following weekend. We hurriedly arranged for my mother-in-law to babysit for us and, late on Friday after I had finished work, we set off in the train from Waterloo to Southampton. From there we took a bus to Romsey, alighting at Ashfield, several stops from the town. Here our lack of knowledge of country life made itself felt! The bus drew away from us, taking with it its pool of light and leaving us in the thick velvet blackness of a country road without any street lights. Of course we hadn't thought to carry a torch, so there we were at the bus stop with about three-quarters of a mile to walk and completely unable to see a hand in front of our faces. Even by crouching I was unable to distinguish sky from hedgerows. We held hands like children and set off in the direction of my sister's house. We felt tarmac underfoot so were reasonably sure that we were on the road, but we had completely lost our sense of direction. Eventually I struck a match, only to discover that we were actually standing on the white line in the middle of the road. We dissolved into hysterical giggles and then to our great relief the glow of a torch appeared in the distance. It was our brother-in-law, Len, who had realised that ignorant townies would not think to bring a torch, and had come to find us. I am sure that if he hadn't found us we would have been wandering around until daybreak and goodness knows where we should have finished up.

The following day I presented myself to Commander North at the Estate Office for my interview. I was asked all the usual questions and felt that my lack of fishing knowledge was a bit of a black mark, although Walter Geary thought it was almost an advantage as I would come to the job with an open mind and with no pre-conceived ideas. Apparently there was one other applicant for the position who had previous experience as a river-keeper, so I returned to London without much hope of success.

Just before Christmas I received a letter from Commander North offering me the job of "Walter's second man" and could I start on 17 January, the first day of the salmon season? It was all very exciting and I hurriedly replied, accepting. There was an awful lot to do in a short time, packing and arranging to move our furniture to Romsey. The Estate would pay moving expenses and supply accommodation for us in Romsey. We made enquiries to several London removal firms but all their prices seemed very high, so Commander North said he would do something from his end.

Marie and I had one more quick visit to Broadlands to see the house we were being given. It was a nice little house on the Romsey by-pass, 4, South Front. We were a little concerned as there was an unfenced stream which was quite deep and fast-flowing right in front of the house, and we had two small children who were liable to fall in and be swept through the hatches immediately below. However, we thought, beggars can't be choosers, and we resolved to keep a tight rein on our girls.

We went home in high spirits, making plans for our great move to Romsey. We decided the best idea would be for Marie and the girls to travel down by train the day before the removal van and stay with my sister, leaving me to come down with the furniture. We packed all our effects into cases and boxes, and Marie and the children left for Romsey on the train. The removal van promised by Commander North was due to arrive at about ten o'clock the following morning and I was ready and waiting. It was right on time, but what a vehicle! Parked outside was a hugh cattle lorry, resplendent with newly varnished wooden sides and an enormous tail-board which let down, normally for cows to embark but now ready for the Aldrich family's home to be loaded up.

We lived opposite a very large junior school, and to the many pupils pouring from the gates at lunchtime, a wooden cattle truck was a vehicle seldom if ever seen. It created a lot of interest and our loading of the van soon had an audience of some fifty youngsters, all very noisy and asking questions about the use of such a peculiar vehicle. The crew of the truck were an elderly little man and his young son. This meant that the son and I did most of the heavy humping, whilst "Dad" carried the smaller stuff and answered the admiring audience's questions.

At last all was aboard and with cheers and waves from the pupils of Woodhill School we trundled off on our long journey to Hampshire. It was not a very pleasant trip, as the weather was really dreadful. When we arrived at South Front that evening, it was very dark and sleety rain was pouring down. Everything seemed to be happening. Marie was there to receive us, as were the gas man and the electricity man, both wanting to read meters and turn on. Our furniture had to be carried from the cattle truck up a steep slope, over a bridge, then about sixty yards along an unlit gravel path with a drop of ten feet or so into the stream to one side, before we reached our new home. I will never know how, but by nine o'clock that evening, all our things were transferred from the truck to the house and "Dad and Son" had had a cup of tea, shaken hands and disappeared, leaving an exhausted, bewildered Marie and Bernard sitting amongst their "home", most of which was stacked in the wrong rooms. By midnight we had sorted outselves into some semblance of order and we retired to my sister's house for a well-earned meal and sleep. We had arrived in Romsey.

CHAPTER TWO

OUR NEW HOME was about two and a half miles away from the main fishing hut which is the meeting-place for the fishermen at Broadlands. Having no form of transport, the first thing I had to do was buy a bicycle. The bike had to be a good one and very strong, as all the roads through the Estate were gravel and hard on tyres. I went up into the town to Davidson's the cycle shop where, for the sum of ten pounds, I purchased a green Raleigh cycle with three-speed gears and dynamo lighting. It was heavy but very well made.

Romsey in 1956 was a small, thriving, market town with a population of around 6,000. It was lovely to walk up the Hundred, the main street, where the people all seemed to know each other and had lots of time to stand and chat. They had a ready smile and a greeting, even for me, a complete stranger. It was very refreshing after our life in London, where one could travel on the same train day after day with the same people and seldom exchange a word.

The Hundred in those days was a splendid street, very busy and usually cluttered with cars parked on both sides, with the bus station in the main square. There were frequent traffic jams but no-one worried greatly. They just waited with good humour for things to sort themselves out. Lining the streets were some of the nicest shops I have ever known: old-fashioned family businesses, still owned and run by members of the family who knew everyone and all about their lives and business. Coming from London, I at first thought this was sheer nosiness, but I soon discovered that it was a genuine friendly interest.

There was Purchase's the grocers, where one was served with care and civility; they even delivered your order free of charge! At Ely's the ironmongers, with its very long, zinc-covered counter and the whole shop smelling of paraffin and tarred rope, you could buy almost anything you required for the farm. If they didn't have what you wanted, the shop down the road did. This shop was my particular favourite. It was a long-established general store named Herbert's, quite large and crammed full with all the artifacts of country life. When you went in there to buy, it could take quite a long time before you were served as the assistant would be talking to another customer, so you waited or more usually joined in the conversation. There was no hurry, but we city-dwellers found a certain amount of difficulty in winding down to the slower pace of life. We were still busy in our house laying carpets and so on, and I despatched Marie to Herbert's to buy stair-carpet clips. She returned with two large boxes of them and an amazing story of trust. Having asked for the clips she was shown two types, and didn't know which one to buy. "Take them all," said the assistant, "and bring back the ones you don't use." He did not know her, but he refused to take any money and let her take the clips, which were quite expensive. We were quite shaken by the trust shown by the shop assistant. It would never have happened in Woolwich!

Just down the road from Herbert's was a beautiful timbered building from where came the most lovely fruity smells. It was the jam factory, and one knew which fruit was in season by the smells coming from the factory. Romsey is duller for its passing.

After a couple of days we were more or less settled in our new home and I was looking forward to starting work the next morning. I then received a summons from the mansion to meet Lord and Lady Mountbatten. I spruced myself up, hopped on my shiny new bike and set off for the "Big House". I felt as nervous as a kitten as I approached the very imposing drive into Broadlands. I had never before spoken to a Lord or, come to that, a "Sir", so I was wondering what to say. I had been briefed to use the term "My Lord" when addressing him. He was, of course, my wartime hero. I knew of his many acts of daring and bravery at sea, and had indeed sailed in a convoy from Gibraltar to Liverpool with Lord Louis as Commodore, but that was the closest I had ever been to him. Nowadays, almost every ex-sailor I meet claims to have been on a ship with him at some time or another and to have known him personally.

I rode up to the front door of the house, propped my bike against one of the colonnades and rang the door bell. Poor soul, I didn't know any better. After what seemed an age I heard shuffling footsteps and the sounds of bolts being drawn, keys being turned and the rattle of chains. I had to suppress my nervous giggles, for it reminded me of "En" and "Min" opening a door during an episode of the "Goon Show". Finally the double doors swung open and I was confronted by a little wizened man in a navy-blue uniform similar to a battledress, with a gold embroidered badge on the pocket, the coat of arms of Mountbatten of Burma.

I don't know who was more surprised, him or me. He asked me what I wanted and when I explained who I was, I thought he would explode! He gave me a royal telling-off for making him come all the way up from his pantry to unlock the front door. I should of course have gone round the back door. When he had calmed down a bit he introduced himself as Frank, the butler. I was soon to discover that Frank was a truly marvellous character and almost one of the treasures of Broadlands.

Frank showed me into the study and there, sitting on a

white leather sofa, were Lord and Lady Mountbatten. I was given a glass of sherry while they shared one. There was certainly no need for me to have been nervous, as they were absolutely charming and put me at ease immediately. We chatted about my past life and especially my sea-going experiences. I reminded His Lordship about the last convoy from Gibraltar when he was Commodore, but he needed no reminder as he had a fantastic memory for detail. He named virtually every ship in the convoy and even remembered the ship which had given the Commodore trouble as she was a straggler and kept falling astern. After about half an hour they shook my hand and welcomed me to Broadlands. I couldn't have been greeted in a warmer manner in my new employment.

Next day was 17 January 1956, the starting date of the salmon season. Walter and I cycled to the fishing hut, opened up, swept it out and lit the fire. Walter had brought his spinning rod down with him and showed me how to assemble it. Then we waited for the first fishermen of the year to arrive. They came at nine o'clock and Walt and I chatted with them about the prospects for the day while we put up their rods. We then walked upstream and they began to fish. The river was quite high but clear and with his second cast one fisherman cried, "Yes!" His rod bent and he was into a fish. The fish seemed to lie doggo, the tip of the rod just jigging up and down "Bloody old kelt," Walter said, and it was. He very quickly netted the fish, unhooked it and gently returned it to the river.

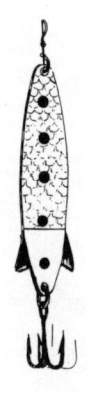

He explained to me what a kelt was and the difference between it and a fresh-run spring fish. A kelt is a "spent fish", one that came into the river the previous season, spawned and is now on its way back to the sea, there to feed avidly and, we hope, return to our river in a year or two as an even bigger fish. The kelts on the Test seem to mend very quickly, but they are still protected by law and must be returned to the river unharmed. Although long and very thin, they lost the red colour of their spawning coats and went an aluminium shade of silver, as opposed to the shining silver of a fresh-run fish (a clean fish straight in from the sea, running upstream, eventually to spawn early the following year). Consequently, it could be quite difficult to tell the difference without a lot of experience. One day Walter was fishing below Longbridge with me in attendance when he hooked a very big kelt. With his rod bent double, he was giving it a lot of "stick" as he wanted to bring it to the bank as quickly as possible. A car stopped on the bridge and four American army officers got out and asked if they could watch the fish being landed, promising not to get in the way.

Walter did not mind them watching, and continued to play the fish. Eventually, I slipped the net under the fish and hauled it up the bank. The kelt was huge, about three and a half feet long and very well-mended. Walt removed the hook and returned the fish to the river. "Gee," said one of the officers, "what did you put a fish like that back for?"

Dear old Walt looked up at him and muttered, "It's undersize." The Americans were amazed and went away thinking that Test salmon must be the biggest in the world.

Such was the humour of my guide and mentor, Walt Geary. He was a super character. Large in diameter, with a ruddy complexion and a delightful sense of humour, in all the years I knew him I never once saw him lose his temper. He was a true "nature's gentleman", with a wealth of knowledge of the countryside and a lot of years of experience on the Test. He had great patience and tolerance with "learner fishers" and gave freely his help and advice. A very happy man was Walt. He had a large appetite and enjoyed his food immensely. He also smoked like a chimney, about sixty cigarettes a day. I used to go to his house at about eight in the morning to find him leaning on the mantelpiece coughing and spluttering, his face scarlet and eyes watering, still with a cigarette hanging from his mouth. "Bloody fags," he would say, "no good to anyone." Then he would light yet another!

To see him fish a fly was an education. He used a thirteen-foot, built-cane Leonard rod made in America, and he made it look too easy, because to him it was. The line would go out as straight as a gun barrel with the fly turning over to perfection and alighting on the water quietly, exactly where he wanted. I couldn't wait to have a go myself. Walt thought it best that I should first learn how to use a spinning rod because early in the year the river was very high and it wasn't really worth fishing a fly. The best chance of a fish was spinning, "getting down to 'em".

I had no fishing gear of my own, so Walter lent me an old rod with a small Pfleuger multiplier reel. He took me up the river to a clear piece of bank and gave me my first casting lesson. This is where his patience was sorely tried. The old reel I was using had no overrun prevention drag on it, so I was continually getting the reel into large birds' nests of tangles, caused by me using too much energy and not putting my thumb on the reel-drum to stop it when the bait hit the water. Consequently a lot of my lesson was spent sitting on the bank, untangling the line. While we sorted out the mess we talked, or rather Walter did, and I listened and learned. He pointed out to me where the fish would lie in the particular heights of water, and why they chose the various places. It was absolutely fascinating. I began to think that Walter had been a salmon in a previous life. He thought like a fish and knew to a foot where they would be.

I was shown how the water, flowing round breakwaters or groynes which were installed at various angles from the bank, create back eddies or small whirlpools, and make an attractive place for fish to lie; the slight contra-flow means that the fish do not have to work hard to maintain their position in the river by swimming against the full current. The same thing applies to almost any large rock or indentation in the river bed where the flow rate is broken sufficiently to cause these back eddies, at the same time providing aeration with available oxygen for the fish.

After a while I was considered to be proficient enough to be left alone to practise. "You're getting it out there and that's the main thing," he said. "Don't reel in, just let it come across with the current, but keep in touch with your bait. Keep the rod down low on the water, so you can strike hard and upwards if you feel a fish take."

"How will I know if a fish takes?" I asked.

"You won't be left in any doubt," he replied. "These spring fish usually give you a hell of a solid pull."

In those days our fishing was let by the day, which meant

that sometimes early in the season there were no fishermen on some beats. At those times we would walk the river in the mornings, with Walter showing me the pools and runs where fish lie, and pointing out one or two of the secret places where he knew he could get a fish quickly should Her Ladyship require one for "The House". In the afternoon he would retire to the sofa in the fishing hut for a snooze and I was sent out on the river to practise fishing. And practise I did! By the end of January I thought I was getting pretty good with the old spinning outfit I had borrowed. I longed to fish one of the "good spots" that Walter had pointed out to me. After Walt had retired for his siesta on 26 February 1956, a date indelibly printed in my mind, I set off with my rod, up onto our number 2 beat.

The man who had rented it could not come for some reason or other, and had telephoned to ask if Walt could give his beat a "run through". Walt said his understudy would, so there I was with a whole beat to fish, all to myself. I had been paying great attention to all the fishing talk over tea in the fishing hut and had learned that a golden sprat was the best bait to use for early fish. Armed with three four-inch sprats and an old Norwegian kidney spoon, I rushed off up the river to Gate Pool, a good spot for springers, Walt had told me. With trembling fingers I mounted the first of my sprats on a sprat tackle borrowed from my "guv'nor", lashed it on with copper wire, added an ounce of lead and hurled it across the pool. I wasn't used to "fishing properly" so it meant that with all the extra weight on my terminal tackle I had several overruns and I could hardly bear the wasted time I had to spend removing the birds' nests. Somehow I did make a reasonable cast and the sprat landed in the water, well across the pool. I carried out my instructions, kept the rod well down and kept in touch with the bait as it drifted slowly across.

My heart stopped as the rod gave a great jerk and before I knew it, a fish was on. There was no question of me striking; the fish, which felt very heavy, had taken off upstream with a very strong, steady run.

I was shaking with excitement, and after about three minutes I saw the fish come to the surface. To my rather inexperienced eye it looked enormous! As I slowly pulled the fish towards me I remembered Walter's teachings: "Don't gaff a fish early on unless you're positive it's a good 'un", or in other words not a kelt. I was sure my fish *was* a good 'un but I just couldn't take a chance of blotting my copybook in Walt's eyes. So when the chance came, I hauled the fish in, bent down and grasped his tail to drag him out.

To me it was the most beautiful thing I had ever seen. But I still had by guv'nor's warning ringing in my ears, so I removed the hook from the fish's mouth and, clutching the fish to my bosom, ran all of a quarter of a mile back to the hut to get Walt's opinion of my splendid fish and have honours piled on my head.

Walter didn't take very kindly to being so rudely awakened and when I showed him my prize he laughed and said it was a well-mended kelt. I couldn't believe it and I nearly cried as he slipped it back into the river. Slowly I walked back to Gate Pool to retrieve my rod. Once there, I removed the very mangled sprat and replaced it with a nice shiny new one. I rather disconsolately flung it out to midstream and allowed it to sink before starting to retrieve. The bait didn't move! I pulled hard and felt a slightly live resistance, then the line began to move slowly upstream with a jigging, up and down motion, similar to the several kelts I had watched other fishermen play. I decided to follow Walter's example and gave him some stick. As the fish came round close to the bank, I dropped my rod and clutching the fish by the tail, very unceremoniously dragged him out. He looked considerably bigger than my first effort, so once again I removed the hook and, cuddling the fish to me, ran back to the hut.

Poor old Walt wasn't too happy to be disturbed twice in an afternoon. "Another bloody gert kelt," he said. He looked at my face, which must have been a study, as I began to walk towards the river to put the kelt back. "Come on, lad," he said, "that fish'll do. Her's a good 'un." He banged the fish on the head and took it into the hut to be weighed. The scales showed the fish to be 19 lbs, and Walt pointed out a patch of several sea-lice, showing that the fish was only a short time out of the sea. A kelt would not, of course, have had any sea-lice, but I was then too inexperienced to look for, or even recognise, sea-lice or to distinguish a kelt from a fresh-run fish. Walter was pulling my leg when he said my fish was a kelt. Marie tells me that I didn't stop smiling for a week!

The next job was to contact the owner of the beat where I had killed the fish. The rule at Broadlands was, and still is, that if a person rents a beat for a day, any fish killed on that beat on that day is his property, even if he does not come to the river. Walter referred to his list of fishermen and found the beat had been leased to Mr Len Heath, a man I had never met. When Walter phoned him he was amazed and delighted that there was a cracking 19 lb fish waiting for him in the fishing hut. He was doubly pleased that it was my first fish.

Mr Heath said he would come down early the next morning to collect "his" fish. He arrived at ten o'clock and duly admired the fish, and I had to go through the whole story of how I had caught it, blow by blow. The fish was gutted and gilled on the steps down to the river alongside the hut and to my delight he cut off a generous portion for Marie and me. From that time Len Heath and I became very firm friends. I was to learn that he had been a great man in the motor-cycling world, having won the Isle of Man TT race on one or two occasions. He was a very nice man and a great fisherman. He not only fished on the Test but also on the Hampshire Avon. He invited me to fish with him on the Somerley water, where he had taken a rod. The Avon is a much bigger river than the Test and I must say I felt a bit lost on such a vast water. I did, however, enjoy fishing a new water and learned a little bit more about salmon lies.

In common with other salmon fisheries, Walter would send off a weekly report on catches, river and weather conditions to the now sadly defunct *Fishing Gazette*, which was a truly splendid fishing magazine. Imagine my pride when I saw in the next week's edition: "Test at Broadlands, 26th February 1956. Three fish killed this week, 18 lb Captain Nimitz [he was the son of a famous American wartime admiral, fishing as His Lordship's guest], 18 lb

Lord Brabourne [Lord Mountbatten's son-in-law], 19 lb B. Aldrich." I felt I had reached the pinnacle of success to see my name in print with such exalted company, and to have beaten them both by one whole pound in weight!

CHAPTER THREE

BROADLANDS ESTATE owns approximately five and a half miles of the main river Test, and apart from one and a half miles of single-bank above Romsey, the remainder is double-bank fishing. The river is quite wide, averaging fifty feet, and in many places very deep. It is a beautiful water, winding down through some very lovely countryside on the Estate. There are long runs and many pools where the salmon lie. Much of the river is good fly water, with plenty of places to spin or dangle a prawn. There were five beats in 1956, each beat being about three-quarters of a mile in length. Then, the stretch in front of Broadlands House was kept as the private Home beat, primarily for trout fishing. The piece of mainly single-bank above Romsey was let as part of the lease of Spursholt House and Mr Morris, who had the lease on the house, fished that stretch with his guests. He was also responsible for the weed-cutting and bank-trimming. Consequently, we had very little to do with that section of our water.

Our bottom beat, the Grove, was let to a Major Magniac, who also leased a large house called Grove Place. This water was often fished by Major John Ashley-Cooper, who killed a lot of salmon there. The top beat, Rookery, was let full-time to Mr T. Instone, the airline pioneer, leaving only three beats to be let to fishermen on a casual basis.

At that time it was possible to book two or three consecutive days' fishing by just a telephone call. The prices varied during the season from ten shillings a day during January up to three pounds for June and down again to fifteen shillings in September. The fluctuating prices were due to the runs of fish and their timing. In those days we opened our season on 17 January and on one very splendid occasion a local doctor caught a beautiful fresh-run salmon of 17 lbs on our first day. It was snowing hard and I had taken another fisherman up to our beat. We were returning to the fishing hut for lunch and to warm up. As we passed the doctor, he signalled that he had got one. We did not dare to stop the car on the snow-covered grass, as we would never have got it rolling again – we had already had a problem in moving it when we left our beat. During the rest of the drive to the hut we were laughing, convinced the poor fool had killed a kelt, for nobody ever caught a fresh fish this early and under such awful conditions. I could hardly wait to tell Walter and to see his face. We opened the hut door to see a fine 17-pounder covered in sea-lice lying on the floor. It was our faces that were a study! But we soon recovered from the shock, and were very happy for the doctor and to "wet the fish's head".

The salmon runs of the 'fifties were totally different from today's. We had a spring run of large fish which came in through January, February and March. These were beautiful fish of 16 lbs and up to 36 lbs. In my first year at Broadlands the first twenty-one fish averaged 21 lbs; in fact, one was considered a trifle unlucky to bring in a fish weighing less than 18 lbs. During April and May and into June we had a run of what we called our summer fish, in the range of 8 to 12 lbs. Finally, in June and July there was a small run of grilse. June was always reckoned to be our best month, with the river full of fish of all sizes, most of them fresh-run with a few of the early big fish still lurking and in reasonable condition. From then on there were not many fresh fish running in, so the fish already in the river became "potted" and after a while very red and the hen fish full of spawn. These fish were considered a bit inferior, so the price for a day's fishing was greatly reduced.

In my early days on the Test we fished for the springers with what today would be considered very heavy tackle. The rods were of built cane, some with steel centres, and the usual length of these rods was nine feet six inches. There was, of course, no fibre glass at that time. Our reels were either centre-pins like the Allcocks Aerial and Hardy Silex, or what would now be thought of as fairly primitive multipliers by Pfleuger or Hardy. The only so called "open-face" reel used was the forerunner of the modern fixed-spool reel, the Malloch. It was a good reel but fairly clumsy and heavy. Hardy's did produce a fixed-spool reel at the time but most of my fishermen were afraid that, although it was beautifully engineered, it was rather light for salmon. As for Walter's thoughts on this modern type of reel, his kindest remark was that they looked like a blooming hurdy-gurdy!

I was taught to use a centre-pin and my first spinning reel was an Aerial, a very simple, nicely balanced reel that needed only a gentle spin to set it turning, and it ran as smoothly as silk. I was fortunate enough to buy at an auction sale a 3¾ inch Hardy Super Silex in "as new" condition. I think that this reel is the ultimate in spinning reels and I use nothing else. Of course I know all the arguments against centre-pins for spinning, the main one being that they are difficult to learn to use, but once mastered they are superb, and for playing a fish there is nothing in my opinion to beat them; it's almost the same as having a fish on fly tackle because the reel is positioned on the rod exactly the same as a fly reel, which allows the forefinger to be kept on the line all the while the fish is fighting. In this way one can almost feel what the fish's next move will be.

The spinning baits we used for early-season fishing were

few, mainly sprats, golden or silver, a Heddon plug or a vibro spoon. The sprats were of the natural variety and during the close season we bought a couple of pounds of them. They would be spread out on the fishing hut table to be sorted in heaps into their various sizes from six inches downwards. Each pile was sub-divided in two; one lot we dyed with Aquaflavine to give them the smoked, golden colour, the rest we left natural silver. When ready, we packed them into Kilner jars in a watered-down formalin solution. I always preferred the golden sprat, mounted on a leaded spinning mount. The mouth of the sprat was slit with a knife or scissors, following the natural line of the mouth, and the leaded needle passed down through the fish until the spinner's vane was in the enlarged mouth. The hooks were laid alongside, with the lower hook by the tail, and bound on with a length of copper wire. They looked most attractive in the water, flashing as they spun, and they did get fish. In fact there was one of my fishermen who fished sprat all through the season, using very small ones of two inches in high summer, and killed his fair share of salmon on them. We seldom if ever see them today, but I am sure a sprat would still catch fish if it was used.

The Heddon plugs were of American manufacture, beautifully made and finished. The balance was critical, and if a hook had to be renewed, one of similar size and weight had to be used or the plug would not wobble properly. There was a British-made plug called a Snapdragon but this one did not work half as well as the American original. Best was the yellow one, but we did lose a lot of fish on plugs, usually just as they were ready to be gaffed. Walt always said that because the plugs were so long and rigid, the hooked fish had a lot of leverage and, by the time it had been played for ten minutes or so, managed to throw the hook, leaving the angler angry, frustrated and sometimes in tears.

Some years ago I was given a plug by one of my fishermen. It was a Heddon and coloured red, with a clear plastic section. Inset through the centre of the body was a piece of gold metal. This plug was thought to be no good as it wasn't yellow but, although it was two inches in length, I found that even in the gin-clear water of summer I caught fish on it. The most effective time to use it, I discovered, was when I had fished a prawn and a fish was plucking at the bait as they sometimes do, but not connecting with the hook. I would remove the prawn and put on my red plug and most times it would be taken with a bang! I don't know why it had such a magical effect, but perhaps the much more violent action of the plug angered the fish and caused them to attack the bait viciously. That old plug has killed many fish and been re-hooked countless times. It finally cracked up and filled with water when used, so I retired the battered and scarred body and it now hangs in a place of honour on my wall. Unfortunately my red plug is irreplaceable as Heddons no longer make that particular pattern. The vibro spoons are also out of production, having been superseded by the Mepps, which is a lighter and better spinning spoon.

At the time I am talking about, nylon was a new discovery and rather suspect, so the spinning lines we used were mostly made of braided silk. The traces were usually of Elasticum wire, which needed great care in use and frequent inspection for kinks, as it was easy to throw a loop when casting; if left without straightening, it could break like a piece of cotton should a fish take.

Our fly lines were also braided silk with a dressing on that had to be greased to make them float. That was the first job we had to do in the morning: run out the line between two fences and apply the grease with a chamois leather pad, then rub it well in with the fingers. At lunchtime the line was strung out to dry and the whole procedure gone through again. It was a bit of a chore, but worth it as those old lines were a joy to cast and one seemed to be able to put them out better in a wind, probably because they were much heavier than the modern lines.

Our casts (I can never get used to calling them leaders) were always made from gut. This material was stiff and brittle and had to be dampened before it could be used. We all had cast-dampers, small round boxes with several pads of chamois leather inside which were soaked in water and the casts laid between them. After a few hours they would then be supple enough to knot a fly on. The casts were made up of short lengths of gut, tied together with a blood-knot every nine or ten inches and quite thick in diameter. The way the casts were made enabled varying breaking-strains of gut to be used, thus giving a tapered effect. For the comparable breaking strain of nylon, gut was probably two or three times thicker, but we still caught fish. So I do sometimes wonder, when I read fishing writers recommending anglers to fish really fine for salmon, whether it is such a good thing. I never use less than 9-lb breaking-strain for salmon fishing. I've seen far too many fish lost by anglers using casts that are much too light, and I do hate to see fish lost with tackle left in their mouths.

When the leases of our Top and Lower beats ran out, the Estate took them back in hand, so in the mid-1950s we had five beats to let. It was at that time that we went over to letting on a seasonal basis Monday to Thursday, not fishing at all on Friday to rest the river, then still letting Saturday and Sunday by the day.

Most fishermen of that period appeared to have an almost casual approach to their sport. By that, I don't mean that they were not enthusiastic about fishing, but it just seemed that there was more time to enjoy the ancillary things that go with fishing. It may be that the majority of the people fishing then were getting on in years and couldn't or didn't feel the need to rush about. All the same, we did have a few younger men fishing then, two of whom taught me a great deal about the art of prawn fishing. They were, I think, two of the best fishermen I have ever known. By "best" I don't mean just at killing fish, for I believe that the number of fish killed by a person is not necessarily a guide to how good a fisherman he is – it may be that he is lucky enough to have the time and money to fish rivers all over the country and be in the best places at the best times. I feel that a really good fisherman is the one who will get fish when other anglers aren't catching them. There is most certainly an element of luck in angling, like weather and water conditions and having a run of fish in the river, but given these, then skill and knowledge of the river really come to the fore, both fishermen who taught me so much knew the Broadlands water intimately, having coarse-fished it since they were lads.

Dick Haston owned a local general store and Fred Jewell was a farmer. Both excellent fly fishermen, they really excelled at prawn fishing. They were most particular about obtaining good English prawns as fresh as possible, firm and red. Their mounts were the Berrie type, at the time an innovation as the only prawn mounts in general use then were spinning or sink-and-draw. Infinite trouble was taken to mount the prawns. The spear of the Berrie mount had a small, flat vane on the front, which made the prawn wobble enticingly, but the prawn had to be very straight or else it became unbalanced and when drawn through the water went round in circles and didn't look very "fishy". The hook part of the mount had small red beads threaded onto it and, when laid along the underside of the prawn, they resembled the eggs or spawn of the female. They looked most attractive when properly tied on.

It was Dick and a guest of his who were responsible for one of my most exciting morning's fishing, and I never even held a rod! They arrived at the hut around nine in the morning and with his usual care, Dick mounted prawns for himself and his friend. I went with them onto our number 1 beat, which is considered a very good holding water for salmon. Dick went up to Black Dog, his favourite pool, and his guest fished Kendle's Corner, a long, inviting-looking bend in the river.

Within minutes they were both into fish, and I was rushing back and forth between them with the gaff. It appeared that wherever they dropped a prawn into the river a fish got hold of it, for almost as many fish were lost as those actually landed. By midday there were eleven salmon on the bank.

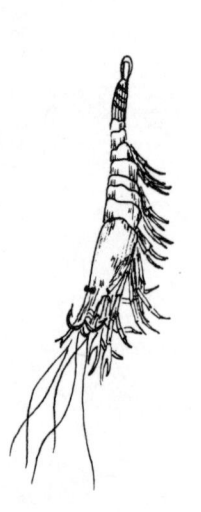

The excitement was so intense that our hands were shaking almost too much to remount the prawns. Anyway, we were all three physically and mentally exhausted. We decided to have a quick lunch and recover our equilibrium a bit and then go out to kill another eleven fish. After lunch we hurried back to the river and they began fishing again. Not a single touch! They tried all afternoon without seeing or feeling another fish. It was as if someone had thrown a switch and the river was dead. It was even more astonishing to learn that although all the other beats had been fished the whole time, there had been no other fish caught. It looked as though there had been a shoal of salmon coming into the river and they had all run up and stopped on one beat. The fish killed averaged 12 lbs and either carried or had just lost sea-lice. The photograph of the eleven fish hangs in our fishing hut as a record of that truly remakable morning for us all.

Dick's usual fishing companion, Fred Jewell, was also an expert fisherman, but in a different way from Dick. Fred, a sufferer from glaucoma, was blind in one eye and only had a quarter "tunnel" vision in the other. He really was a super chap, with great patience and infinite gentleness and kindness. Because his sight was so bad he fished by feel and judgement, using his knowledge of the river. He seldom wore wellingtons, preferring thin, leather-soled shoes. I can see him now, sliding his feet along a fishing platform, feeling the joins in the planks until he was in the right spot, then pulling off line in arm-lengths until he had enough to drop the prawn into the taking place. "That's about it," he would say, and in a while a fish would take. I spent a lot of time with Fred, helping him by gaffing his fish or assisting him over rough ground. Just watching him and listening to his little commentary taught me so much about "feeling the river".

Both Fred and Dick showed a great interest in the river apart from the fishing. Dick would often show up in the evening and work with a will at weed-cutting or any of the jobs we were doing in or out of the salmon season. Fred helped in different ways; for instance, it was only through him lending us his tractor, grader and heavy roller, complete with a driver, that we now have a made-up track to the upper beats. The Estate supplied the gravel and we carted it, but all the rest of the work was done by Fred. He became a very dear friend whom I sorely miss now he is dead, but his memory and teaching live on in my mind as I fish some of his favourite pools.

The river at that time was totally different in character

from the Test of today. For instance, in January almost every year the river would be in full flood. It was such a regular occurrence that the folks living near the river were always prepared for it. A few hundred yards from our main fishing hut were a pair of beautiful-looking thatched cottages, almost built for the photograph on a chocolate box. Early every year, the tenants of these houses would fetch from their sheds a series of trestles and planks and erect a long raised platform which they had to use to reach the road during times of high water. The river, although bank-high, cleared very quickly and took on a lovely translucent green. The water was certainly clear enough to be able to see a golden sprat spinning some four feet deep.

The life in the river has changed, too, over the years. There were crayfish in abundance, which Walter loved. Sometimes after tea I would be dispatched upstream with a bucket to get some for his dinner. It was a very simple matter to fill the pail in an hour with crayfish, some four inches long. Under almost every large stone or in the roots of weed were crayfish which needed only a quick snatch to be caught.

We have always had, and still have, the most beautiful flora growing on our river banks, with plants like musk, or at least that is what we called it. A variety of wild orchid, this lovely flame- and yellow-coloured flower only seems to thrive by the river. I have tried to grow roots taken from the banks of the Test in my garden. They all failed dismally. Teasels and bullrushes grew in abundance and still do.

It is the "in-river" conditions that are changing so dramatically. Although the river was almost over-keepered mile for mile in those days, there was always plenty of wildlife in and alongside it. The coarse fish like dace, roach, chub and grayling were regarded as vermin. Pike in particular were to be killed at all costs. Walter killed more than two hundred pike in his first four years at Broadlands, many of them on the now frowned-on trimmers. This killing method consisted of a live bait attached to a large cork or float with a long length of heavy line tied to a very flexible willow wand. The bait was left to swim around in a piece of dead water, the sort of place pike inhabit. The pike would then take and gorge the bait and play itself on the willow wand, to be removed by the keeper when he came along.

On one occasion in 1920, a large pike was seen by Walter, while walking with Colonel Ashley, lying off the lawn in front of Broadlands House. (Colonel Ashley was Lady Mountbatten's father, from whom she inherited Broadlands.) The fish was too far from the bank to snatch or snare, so Walt was sent back to the house for the Colonel's 12-bore shotgun. The pike remained in its position while

the Colonel took aim and fired both barrels in rapid succession. The fish was peppered with shot and came to the surface, and Walt gaffed him out. It weighed 17 lbs, a good fish not to have in the river.

In my early days at Broadlands Walter taught me the art of snaring pike. It is quite an exciting thing to do. The best time for using a snare is February to March, when the pike are thinking about spawning and come in close to the banks, lying in shallowish water between the weed. Our equipment consisted of four drain-rods which would be screwed together to give a long reach. On the business end was a four-foot length of very stiff wire with a large rabbit snare attached. Once a pike has been spotted, the idea is to lower the open snare very gently into the water about four feet above the fish and slowly bring it downstream over the fish's head. As soon as the snare reaches behind the gill covers, back up the bank, closing the snare and dragging the fish ashore.

I was always amazed how the fish just sat there, waiting for the loop to go over its head. It was a fruitless exercise to attempt to snare pike by putting the snare over its tail first; it just never worked, as the fish seemed to be aware of the game and rapidly swam off. The size of the fish didn't seem to matter as this method was equally successful with small or large fish. One of the first pike I snared looked to be about 10 lbs as all I could see was a head protruding from a weed-bed some eight feet from the bank. I managed to reach the fish by joining all the drain-rods together and eased the snare over the fish's head. The loop of wire must have touched the pike for he suddenly took off, tightening the loop, dragging the rod through my hands and removing quite a lot of skin from my palms before I stopped him and fought him back onto the bank. That fish weighed in at 25 lbs, the biggest pike I have ever caught.

The fact remains that we never really got on top of the problem and the vermin that remained and were caught later were superb specimens. 3-lb grayling, dace of more than 1 lb, roach of 2 lbs and enormous chub of record proportions swam in our river. I remember Fred Jewell hooking a fish one Sunday morning in the salmon close season and calling me to bring a net to remove a large chub he was playing. It turned out to be a roach of 3 lbs 9 oz, which won the *News of the World* rod and reel that week.

On another occasion a coarse fisherman called me, thinking he was into a salmon on a single maggot, as did often happen. I ran round with a big net, waiting to take the fish out. I saw it was a very big chub and told the angler what it was. He literally went to pieces and finally lost the fish of a lifetime. Without any doubt in my mind, the lost fish was a record chub of 8 lbs or more! I am sure there are still chub of record-breaking size in our river. Grayling, too, but far fewer.

The dace and roach do not seem so prolific or of such a size as they were in the 1950s. We had a small group of chaps who fished for salmon and traditionally came to Broadlands on Boxing Day for coarse fishing. One of their number was a delightful man called Skip Reeves, the secretary of the Christchurch Angling Club. On one Boxing Day morning, Skip had sixteen dace, all of more than a pound. Such a catch is seldom, if ever, seen today.

The fly life was seasonal, too. Stone fly came up in March. The grannom hatched in mid-April. They came off in clouds and their upstream migration was a sight to behold. There is one strange thing regarding grannom. While below Romsey they hatch in profusion, above the town they are virtually unseen. My theory for this is that the early April weed-cutting allows the caddis of the fly, which adheres to the weed, to float downstream, thus stocking the lower reaches with an abundance of larvae.

After the grannom there was a lull in fly hatches. Walter always said that the trout got back into condition on the grannom, ready for an enjoyable feast on mayfly. You could set your clock by the mayfly hatch. It began at the end of the first week in May. Two weeks later it would all be over. It is a fact that the mayfly came off the water in such large numbers that they frequently stopped traffic crossing the river over Middlebridge, when windscreens became covered with the bodies of flies.

We still have very good hatches of all the flies mentioned but quite often they come off the water together or overlap, leaving long periods of time with almost no fly life on the river, and consequently blank days with no fish caught.

CHAPTER FOUR

THE HISTORY OF Broadlands salmon fishing is an interesting story in itself. It all began in the year 1880. The agent for the Estate, then owned by Lord Mount Temple, was G. R. Kendle. It appears from the records that he was walking by the river when he saw a salmon roll over. Mr Kendle was a confirmed salmon fisherman, but until that time had only been to Scotland in pursuit of the sport. He went home, returned to the river with his rod and very soon had a fish of 14 lbs on the bank. This obviously set him thinking of the potential asset that six miles of river, hitherto unused and rather neglected, could be to the Estate. So for the rest of the year he fished for salmon on the Broadlands stretch of the Test, and he certainly killed a few fish. Exactly how many isn't recorded, but it must have been enough to convince Lord Mount Temple that it would be worth employing a full-time river-keeper to put the river in order and turn it into some semblance of a salmon fishery.

The first full-time river-keeper was a chap named John

Cragg, who began working for Broadlands Estate in January 1881. From the two or three photographs I've seen of him, he looked a big man, dressed in gaitered trousers, a short-tailed coat and a square, stove-pipe hat. It appears that when he first arrived he could not read or write very well, as all the early summaries in our large record book were written by Mr Kendle. Cragg was a tough man, the scourge of local poachers, who worked hard to clean up the river and its banks and he fished, too, for the first two fish in 1881 were killed by him on 19 and 24 October. There does not seem to have been any close season in those days as fish were killed from October through to the following October.

In that first season forty-six fish were taken by about four rods and the keeper. Mr Kendle accounted for sixteen to his own rod and Cragg had eight. Not a bad showing for a maiden fishery. The second season's fishing was even better, a total of fifty-six salmon being killed. The water was let to a Mr William Clifford on a seven-year lease, allowing three persons to fish. Unfortunately, he died during the first year of his tenancy and the remaining time of the lease was taken over by his two friends, Colonel Leigh and Mr Basil Field. Those two gentlemen, with Mr Kendle, must have had wonderful foresight, determination and energy, for they set about improving the runs of salmon in a manner that we could well take lessons from today.

Their first task was to see that the fish had free access to pass upstream through the then working mills on the Test. They went downstream to talk to the millers and agreed with them a sum of money to ensure that one of their hatches would be kept open for the passage of fish. This accomplished, they turned their attention to saving the young fish, parr and smolts. At that time most of the land in the lower Test valley was given over to water meadows. These fields would be flooded by means of a complex system of hatches and carrier streams running from the main river. It was very cleverly engineered, with carefully calculated levels, a flooding stream and a draining one at a lower level. The meadows would be left covered in water for a few days then one hatch would be opened and another closed. The water would drain away, leaving all the alkaline goodness on the land and giving some of the finest grassland in the country.

John Cragg was greatly concerned about the water meadows and wrote a report to his employer which is worth quoting as it was written, bad spelling and all, for it is full of good recommendations:

The river been so low during the summer months and the Salmon are not able to come up the river and what few do run up will not stop on the lower beats. Owing to the water not been to its proper height and is not fishable and so much complaints with the gentlemen of the water been so low it wood be adviserable to have a regulation on Water Meadows for the summer months. For farmers to have *all* the water from Novr. up to April 1st for watering there meadows and after 1st of April for there sluice atches to be left down from 6 a.m. on Monday morning until 6 p.m. on Saturday night, and then for the farmers to have all the water from 6 p.m. on Saturday night till 6 a.m. on the Monday morning. And I have no dought that if you could come to this arrangement it would greatly improve the fishing on the lower beats and I have been speaking to some expearance farmers and they say that wood be quite sufficant water for any farmer to have and wood get much better crop that way than if he was flooding his meadows night and day. And there should be a proper grating fixed in front of the water meadow atches with small enough mash so as to prevent the young Salmon from drawing down the meadow as they have been dowing. And also to stop kelts from drawing down after been spawning. For when the man slips the water and sometimes changes it from one meadow to another the young of the Salmon and the kelts are left there to die on the field. With regard to the netting at the sea, a thing that ought not never to be allowed in such a small river, as it does not give the upper Proprietors a chance for Salmon fishing and they have all the trouble of looking after the Salmon all winter in Spawning time and there ought to be a Bill brought in by the Government Inspector to stop the netting on a small river like the Test at the mouth of the sea. It also requires a good man to look after the netting at the seas for the Summer. And it requires two men hear at Romsey by night to look after the Salmon during the time they are spawning on the beds. And one man by day.

Having read this very constructive report, the three intrepid gentlemen negotiated with the water meadow farmers to fit the screens recommended by Cragg over their sluices to prevent the small fish being swept through. There is no record of the sums of money they paid for the farmers' and millers' co-operation, but it seems that everyone was happy with the arrangement, and it worked well. We are not told how many fish the estuary netsmen killed, but I imagine that it was a lot. The three gentlemen set out to meet the owners of the netting rights and after a lot of haggling found a price to suit everyone, so all the nets were bought off. The amount of good that this tremendous piece of foresight did for the salmon is beyond calculation.

Although they had already accomplished a great deal to help the salmon, they still had many problems to overcome. For as the runs of fish increased, so did the numbers of smolts going down to the sea. As late as 1893 it is recorded that the eel-traps at Romsey were not observing the close time which prohibited them working between 1 December and 24 June, and at those times they caught a great number of smolts which were openly on sale in Romsey at 8d per pound! There was also dreadful pollution by the town sewage and by refuse from the paper works and tannery yards, all of which did a lot of harm to the spawn and young salmon. As there was still no Board of Conservators, fighting the pollution was difficult, but the lack of such a board did have a bright side, for the Test was probably the only salmon river in England allowing anglers to fish without a rod licence.

I just thank Providence that our river was in the hands of these three fine men. If only I could have met and got to know them as, with the technology that we have today and their will and determination, I know the salmon would not be suffering as it is now.

In 1883 the same three men undertook a stocking programme that makes the mind boggle even in modern times. That year it is noted that they turned into the water 2,000 yearling "Loch Leven trout", 1,220 "common trout" (whatever they were) and, incredibly, 4,000 American salmon fry. When I think of the problems involved in bringing those fry all the way from the USA, I just don't know how they tackled it. In those days it must have been at least a seven-day journey across the Atlantic, and the task of keeping the fish alive with the then primitive oxygenating equipment and tanks would have daunted most people, however keen. But they did it, and carried on with a similar but increasing stocking policy, for up until 1910 it is recorded that an average of 9,000 trout a year were turned into the Broadlands stretch of the Test and most if not all of them came from other river systems as far afield as Scotland in the north and Devon and Somerset in the south.

So what of the indigenous Test trout and indeed salmon? Is there such an animal? If so many so-called "foreign" fish have been put into the water over the years, there must have been cross-breeding and the pure strain would, I think, have been bred out. It's an interesting theory which I would like to put to those who think that importing "foreign" eggs and fry will damage the genetic strain of fish that are already in the Test (I hesitate to use the terms "natural" or "indigenous").

Also of interest is an item recorded in our book: "Two of the largest Salmon ever taken, and weighed in the fish shop in Bond Street, London, in 1880. Caught in the nets at The Haggis Fishing Bank 2 miles below Newbargh 70 lbs in weight, 4' 5" in length, 2' 7½" in girth, weighed by Mr. Frank Buckland who named it the 'King of Scots'." Several other large salmon are also recorded in Scotland, a cock fish of 68 lbs in 1893 from the Tay, in 1890 63 lbs from the Esk, 62 lbs from the Tay and a 60-pounder from the Severn. Where, oh where, are the fish of that size today?

It was in 1883 that a certain Major Dunn killed *our* record fish on 11 May. It weighed in at 43 lbs, was 3' 10" long and 2' 11½ in girth. What a fish! I very much regret the passing of the big springers on the Test and cannot honestly believe those fishermen who say they would sooner have two fish of 11 lbs than one of 20 lbs. I know which I prefer and I'm *not* one of those who think that big is beautiful and therefore better.

In the meantime Cragg was still beavering away to improve the fishing. I think he was also a likeable rogue. At that time a lot of the farms on the Estate were let out to tenant farmers. One of the biggest was the 500-acre Lee Park Farm, which had quite a long stretch of the salmon water running through it. As a makepeace to the farmer it was tradition that he would receive a salmon to compensate him for having fishermen walking over his land. It was the river-keeper's responsibility to see that the farmer got his fish, but what sort did Cragg send up to the poor old chap? The fish could only have been a well-mended kelt, as can be seen from the following which is a copy of the original letter in the record book:

Dated 1892, From the farmer at Lee to Cragg the River Keeper.
Dear Cragg,
I have sent you back the fish as soon as I could as I do not care for Salmon when it is *white*, or in other words, out of Season. The one we had last year nearly turned me up, this one I am sure would *quite*. If the risk and nuisance of having strange horses in the stables is not worth a clean fish, then all I say is let the Gentlemen go elsewhere.
Signed H. De La Mare (Farmer)

I must say I have a great deal of sympathy with Mr De La Mare, but equally I quite like the rather nasty try-on by Cragg. Strangely enough, I met a lady at a dinner party recently who was in fact related to that very farmer, and when I told her the story she was most interested. She knew many tales of him and it rather looks as if he was a miserable old devil, so maybe John Cragg's hoax was justified after all.

John Cragg remained as river-keeper until 1905. It seems

that during his later years he was ably assisted by his son, who took over from him when he retired. Cragg Junior, as he was known, carried on in his father's tradition until 1910. There is no reason given for his leaving, only a note that a man named Cheyney was appointed as river-keeper at that time. He lasted only three years. He didn't seem to have much impact on either the fishery or the owners, for apart from the note of his starting, nothing else is written of him. In 1913 the legendary Walter Geary came to Broadlands. He told me that he only came on trial and wasn't sure whether he would like it there. He must have decided that he did, for he served for fifty-one years on the river. When he first arrived, he lived in the bothy with several other unmarried lads, and from what he said they had a rare old time together. The old idea of a bothy for a group of single lads was a good one. All it consisted of was a house sleeping four to ten people and one large communal room. They would fend for themselves, apart from one good evening meal cooked by a lady living nearby. This system could well be put to use today for the temporary housing of students or trainees on a large estate.

Some very large brown trout were caught at Broadlands in the late 1800s, big even by today's standards. The reason for their great size is fairly simply explained. Romsey is built over several streams, all running into the main river. One of these streams flowed under the local butcher's shop, who also slaughtered his own beasts. All the unwanted innards of the animals killed were quite naturally thrown into the very convenient stream and were carried down to the main river just below Middlebridge. There in the junction pool the brownies scoffed back the free food, and consequently they put on weight at great speed. In 1884 a trout of 12¾ lbs was taken from this pool on a "lump of liver"; that superb specimen sits in a case in my fishing hut today. There was also one of 19 lbs preserved in a glass case at Broadlands House. Unfortunately, during a boisterous children's Christmas party it was knocked from a shelf and broken into a thousand pieces, but I did see it, and again it was a beautifully shaped fish. That, too, was caught on a piece of meat – whatever happened to the fly purists of yesteryear? I think most fishermen are poachers at heart!

One evening in the early 1970s we were sitting at home when there was a knock at the door. I opened it to reveal a tiny woman in wellingtons and all muffled up against the cold. She asked if I was the river-keeper and when I confirmed that I was, she introduced herself as Miss Amy Kendle, daughter of G. R. Kendle, the founder of Broadlands salmon fishing. She was then in her late eighties and a very remarkable lady – bright as a button, nimble and with a very clear memory. She told us many stories of her young days at Broadlands, with graphic descriptions of the houses and how they had changed, and of her trips to the river with her father to catch the elusive salmon. We visited her at the Edwina Mountbatten Memorial Home, where she produced many early photographs which I found really fascinating. One very large print which she presented to the Estate Office, where it still hangs, shows the Estate workers of 1890 and a pretty sinister-looking lot they were with their full shaggy beards and large flat caps, the hierarchy foremen and coachmen wearing bowler hats. It would be a good idea if the Estate workers of today had a similar photograph taken for posterity to show what an equally evil-looking lot we are!

Miss Kendle was a wonderful person and we loved her dearly. In her ninetieth year she planned a party which we attended, and a good time was had by all. She was already looking forward to her hundredth birthday party. She did have one great wish and that was to plant a tree on the Broadlands Estate to the memory of her parents. There could be no better place for it than where her father killed the first salmon for the Estate. I approached Lord Mountbatten regarding the memorial tree and he readily agreed that it should be done and said that he would provide a plaque to be set up alongside the newly planted tree. He also extended an invitation to the small "planting party" to go back to Broadlands for tea.

The day of the planting ceremony was a very windy one. Miss Kendle, though, was made of very stern stuff and insisted on carrying on with the arrangements. There was no question of me digging the hole; she wanted to do the complete job herself. Well wrapped up against the weather, she duly did the excavating and the small copper beech tree was put in place. Later we erected a tree-guard around it and put in a few crocus and daffodil bulbs and, of course, the plaque to commemorate the day. The tree is thriving and the bulbs give a nice splash of colour to the river banks in the early spring.

Sadly, Miss Kendle did not live to celebrate her one hundreth birthday. I regret never recording some of her stories of the early days at Broadlands, but her tree stands as a fitting memorial to her and her parents on the bend of the river named after her father, Kendle's Corner.

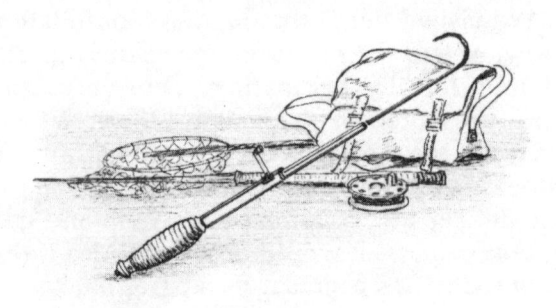

CHAPTER FIVE

AT BROADLANDS there has always been a very friendly atmosphere, with Lord and Lady Mountbatten regarding the Estate personnel almost as part of their family. This family feeling generated in the staff a similar sense of belonging. His Lordship was always available to help with any problem, either to do with the job or personal. Every Christmas there was a party at the mansion for the Estate staff and their children, given by Lord and Lady Mountbatten. Our children sat down to tea with members of their family. There was a present for every child, taken from a huge tree set up in the main hall and usually handed out by Lord and Lady Louis. I think they enjoyed these parties as much as we did.

We were invited to all the big occasions taking place. I well remember the wedding of Lady Pamela, Lord and Lady Mountbatten's younger daughter. The wedding took place on 13 January 1960 and it was a white wedding in every sense of the word, for it snowed heavily all day! We had a reception at the Crosfield Hall and it was there that I had my first dealings with the so called "gentlemen of the press". It was a glittering occasion with many VIP's present, and the pressmen behaved very badly as they pushed and shoved the guests out of the way to take their photographs, even though they had been told that the wedding party would give them lots of time and were prepared to pose for them.

It was barely a month later that we were all terribly shocked to hear of the sudden death of our lovely Lady Mountbatten in North Borneo. The whole of Romsey was stunned. The Estate in particular felt her death deeply, as her familiar figure was often seen walking the grounds and she would always stop for a chat with anyone she should meet. She took a keen interest in everyone on the Estate and knew most of us by our Christian names. She was particularly fond of the children, who adored her in return.

On the sad day her body was coming home to Broadlands, I was asked to meet the hearse some miles from Romsey and guide it by way of the back roads to the House, avoiding the many press photographers waiting at the main gates. I was taken in a police car to the rendezvous, met the cortège and duly brought her home. His Lordship was waiting, shook my hand and thanked me. It was a very moving moment. Her body was later taken to Romsey Abbey, where the whole of the Estate staff took two-hour spells to stand watch with her for the time she lay there. That, too, was a moving experience. The number of people filing past was astonishing. They came from all denominations and creeds, from all parts of the world and all classes. Even at four o'clock in the morning, there was still a long queue waiting to pay their respects.

After the funeral the Estate began to recover. There were changes, the main one being that we no longer had a resident agent. The agent's job was taken over by a firm of estate agents. It most certainly wasn't the same, and we all felt the loss of the personal touch. Our Commander North had retired and we did miss him greatly, for he was always there for a chat – he knew us, and we him.

In retrospect, a lot of the feelings were generated by our resistance to a change in the traditional way of life that we had become used to. During this period it was difficult dealing with the new agents. We had problems getting decisions because of the channels they had to go through. Slowly we began to get used to the new system, though I still feel that it doesn't work as well as before. I think an estate the size of Broadlands must have an agent – one man, not a company – to whom we can go for quick action, whether it be to do with the farm, shooting, fishing, forestry or maintenance. On the other hand, the problems where interests conflict are nearly always sorted out by heads of department getting together, another example of the "family feeling" that we must all live together in the closest harmony possible.

The Estate has always been of interest to local people. As Southampton grew in area, with an ever-increasing population and much of the open land being built upon, they looked for green areas for walking. This gave us a lot of trouble with trespassers. I do appreciate that townspeople need to get out and about for walks, picnics and so on. I am also aware that my riverside is an ideal spot for these activities, but I am afraid I am rather cynical regarding the Great British Public. Many of them do in fact respect the country code, but by and large, they tend to litter and spoil any lovely area given over to them. I can quote many instances. On an island by our main fishing hut I planted a few snowdrops and over the past twenty-five years they have spread so that now each spring there is a complete carpet of flowers. In spite of "Private" notices I have seen many seemingly nice people coming with trugs, trowels and gardening gloves, digging up roots to take away. When approached, they usually say they thought the flowers were wild.

Mind you, trespassers do sometimes have a lighter side and were responsible for one of my most embarrassing moments. It happened some twelve years ago in June, and it really was a flaming June with clear skies and the fishing particularly good. On the lower beat there was a very crusty colonel fishing. The old boy was a meticulous fisherman who was very fussy about all things relating to his sport. He would not tolerate anyone stomping along the river bank, so on his day at Broadlands I took great care, when visiting him, to walk well back from the river and also to keep a good eye out for any strangers wandering about. I had taken the colonel's guest down to the bottom of the beat on the left bank, and we were making our way back upstream to the hut for tea when to my horror, I saw about ten people, complete with picnic baskets, on the right bank. They were still a field away but were undoubtedly heading at an angle towards the river bank halfway down the colonel's beat, obviously intent on having tea in that delightful spot.

I left the colonel's friend and rushed up to the bridge, crossed it and hurried downstream in pursuit of the trespassers. Meanwhile, the group had disappeared through a hedge into the next field, but they were still on the old boy's beat. At length I reached the hedge, breathless and sweating freely. Sounds of merriment came from the opposite side and they were clearly having a wonderful time. It did seem a shame that I was about to spoil their fun, but my vision of facing the wrath of the colonel pushed all pity from my mind. I plunged through the hedge in a most undignified way and was confronted by an astonishing sight. There, gambolling about in the long grass, were ten assorted bodies, all nude! There were mums, dads, grannies and grandads, children – all different shapes and sizes.

Red-faced and puffing, I was lost for words. After a moment I did manage to mumble something like, "What's going on here?" They all gathered round me, showing no embarrassment as they asked what was the trouble and told me they weren't doing any harm. I felt strangely over-dressed in slacks and shirt. I singled out a dad and drew him aside, asking for an explanation. It appeared that they were two families of naturists. I told him the situation and said that they could not be allowed to frolic in the sun on my patch of riverside. I was greatly relieved when they dressed and good-naturedly went off to find pastures new. As a parting shot, one of the grandads did suggest that I had the ideal place for a nudist camp. I don't really fancy myself as the keeper of such an establishment – but, if ever the fishing became terribly bad....

We have over the years built up a tradition that our fishermen return to our main fishing hut at four o'clock for tea and to chat over their day's experiences of fish and fishing. When the weather is warm our tea-table is set outside under the chestnut trees growing close to the river, where there is always a cooling breeze. Walter called it our "lumbago corner", as the breeze caught his back through the open chairs and caused him a lot of suffering. One very

warm afternoon, all was ready for tea outside and I had popped into the hut to see if the kettle was boiling. On my return I found a charming elderly couple sitting at the table. "Two teas, please," they said. Naturally I gave them tea, but then explained that we weren't a café and sent them on their way somewhat refreshed.

We are very fortunate in having our main fishing hut built only a few yards from the river on high ground, which means that it doesn't become flooded during high water. The hut was originally put there midway through the war by the RAF. Made from corrugated iron sheets, it is a traditional service nissen hut. Throughout the war seven servicemen lived there, and I would imagine they had quite an enjoyable time of it. Their purpose in being posted to such an isolated spot was to operate a dummy town: a series of flimsy buildings that bore a resemblance to a town when viewed from a high-flying aircraft. The whole thing was built on open land far away from Southampton. The buildings were packed with old tyres and anything that would burn well, so when the Nazi bombers came to destroy Southampton, the dummy town would be set on fire, in the hope of deceiving the raiders into dropping their bombs on the open fields. Rumour has it that the only time the mock town lit up was when it was struck by lightning and it burned all day! I can't think that the plan was a very great success as poor old Southampton suffered a tremendous number of German bombs.

However, the really happy outcome of it all is that we have a very good fishing hut, although we have improved on the original tin one by concreting over the outside, so protecting and preserving the iron sheets. A few rambling roses grow over it, making the structure look more picturesque and fitting in with the surrounding countryside. In Walter's early days his headquarters was a little wooden hut some eight feet by six feet, but I don't think we could manage with one that size today.

It is often said among river-keepers that we don't really know our water until we have fallen in a time or two. I know my water well! I have read and seen on television a great deal of advice about going out of your depth while wearing body-waders none of which comes to mind when it happens to you. One June, while I was weed-cutting, I had just finished a section and was inspecting my work. There was a single streamer of weed left, but I managed to reach it from the bank by hanging on to a branch of a tree and leaning out over the river. The branch broke, dumping me full-length into the water and filling my waders. I wasn't in too much trouble and I was close enough to the bank to grasp a bush. The bother really began when I tried to get out of the river. With about a hundredweight of water in each leg, I found it impossible to lift myself out over the steep bank. It was a ludicrous situation. I was powerless to move. Eventually I was able to reach down inside my waders to find my knife, open it with my teeth and cut a long rip in the lower part of the boots. It was only then that I was able to struggle out of the river.

My most frightening immersion in the Test involved a Land Rover. I was returning to the fishing hut from visiting a fisherman on the lower beat and swung the Land Rover into the hut compound. I put my foot on the brake pedal to stop at the hut door but, to my horror, my foot went straight down to the floor. The vehicle kept going past the hut, over the bank and into the river. It finished up about six yards out, with the freezing cold water up to my armpits. The shock was awful! All the breath was driven from my body. I was very near to panic as I knew the vehicle was on the edge of a twelve-foot-deep hole. Had the car slid into that deep pool, I would have had no chance of survival, for the windows of the Land Rover were far to small to allow me to escape, dressed as I was in heavy weather clothes.

All the learned advice I had read about how to extricate yourself from sunken cars came back to me. If the windows were open to equalise pressure, the doors would open easily. Bearing in mind that the car was at right angles to the bank, the upstream door bore the full brunt of the current, so in theory, the downstream door should have opened fairly easily. But it didn't! I had to brace my body against the upstream side and force the other door open with my feet. Even then it was a devil of a job to move it. However I did manage it, probably with added strength that fear gives in these situations. I stepped down from the car into the twelve-foot-deep hole. I think I was literally only a foot from death.

Getting the Land Rover out of the river was another kettle of fish. At the time of my amphibious Land Rover, we had an army of sea cadets on the river. They came from all over the country for their "crossing the river" championships. This involves great tripods of wooden beams, blocks, tackles, ropes and breeches buoys, lots of noise and excitement, and a great deal of fun. The Land Rover, half immersed in midstream, was a tremendous added attraction and diversion. The cadets were rapidly organised into fetching their blocks and tackles, lashing one end to a large chestnut tree and hitching the other to the tow-bar of the

Land Rover. What seemed like hundreds of youngsters heaved with great enthusiasm and the vehicle rolled gently back to the bank, but no amount of effort could drag the car up and over to the shore. We stopped a passing tractor, hooked a chain onto the Land Rover and with a mighty heave my vehicle returned to terra firma. The cause of the brake failure soon became obvious: the rear wheel was projecting from the side of the car by about a foot, showing that the half-shaft had failed which allowed the brake drum to slide away from the brake shoes, leaving them to expand onto thin air. We drained all the water from the engine and rinsed it out with oil and petrol. To our amazement it started and kept going, which says a lot for the robust stamina of the Land Rover.

Chapter Six

In the contract that I signed for Commander North when I started work at Broadlands was a clause stating that I would be required to assist in any way possible on shooting days taking place during the fishing close season.

I had no experience whatsoever of game shooting, so I was very much in the hands of Bert Tiller, the retiring head gamekeeper, and Reg Blake who was about to take on his job. It was proposed that I would be in charge of a line of beaters, about twenty in number. I was taken round the land we were shooting over and shown how it was to be driven. Some of the woodlands on the Estate were very large and to drive the pheasants out and over the guns successfully meant that sometimes two or three lines of beaters would be needed. Each line started from a different point and eventually linked up, forming one continuous row of men. This was all right in theory but quite often, due to the size and density of the woods, we would miss each other and plaintive cries could be heard coming from one group to another. Beating is hard work but very enjoyable, and we always managed to get it right on the day.

The shoot in those days was a family syndicate and stayed at a fairly low level as far as numbers of birds shot was concerned. When Reg Blake left Broadlands to go to another part of the country, we took on a new head keeper, and so began a twenty-five-year association with Henry Grass, a member of the famous family of keepers going back many generations. Mr Grass, or Harry as we know him, set about improving the shoot in no uncertain way. He brought with him great knowledge and expertise which he used to increase the number of birds reared, and he introduced a new blood-line of pheasant to the area. From that time the bags of birds climbed dramatically.

A story that Lord Louis told me with a certain amount of delight was that Mr Grass would never tell him how many birds he reared for the shoot. "All Grass says is, 'How many birds do you want to shoot in a season, and I'll provide them. If the numbers are not what you expected at the end of the shooting season, you can take it up with me then.' "All I'm allowed to know," His Lordship said, "is how much it all costs and to pay for it!" Lord Mountbatten enjoyed telling his friends that he only owned the shooting but his keeper actually ran it.

I must say here that I do not particularly enjoy shooting – actually holding the gun and pulling the trigger – but I do love the atmosphere, the people involved and everything about the day of the shoot. I am most certainly not anti-shooting.

On our big days the people shooting used two guns, which meant that they had a man with them known as a loader. His function was to stand slightly behind and to one side of the shooter, carrying the second gun; when both shots had been fired he would take the used gun from the shooter, replace it with the loaded one and as quickly as possible reload ready for the next change. There is a considerable art to loading, and above all it must be done safely, smoothly and quickly. I was fortunate enough to be taught how to be a loader by Harry Grass. On my first lesson he pointed out all the mistakes that can occur, like clashing barrels during the change, but all the time impressing on me that safety must come first, for the 12-bore shotgun is a very lethal weapon.

This was demonstrated to me by Lord Mountbatten, who took me to the clay pigeon field where he was to practise before a day's pheasant shooting. He knew I had never handled a 12-bore, and patiently explained the rudiments of using one. He then handed me the gun and told me to shoot at a bale of straw from ten yards' range. This I did, and one barrel blew a very large, ragged hole in the tightly packed bale. "Imagine that bale was a man," His Lordship said.

The sight of that hole in the straw stays indelibly printed in my memory and I can see it whenever I handle a gun. It was the finest lesson I could have had and should be given to all users of guns as a demonstration of what could happen when a 12-bore is carelessly handled.

As the shoot improved, the gamekeepers had to spend more and more time on their ground, looking after the great number of pheasants being reared. Traditionally, the head gamekeeper acts as loader for his employer, whether shooting at home or as a guest on other shoots. But to help Harry Grass, I was asked to take over the loading for Lord Mountbatten, enabling Harry to remain on the Estate to do the many jobs entailed in running a shoot the size of Broadlands. It was a great privilege for me to be with His Lordship and to travel with him as his loader and driver. He was a very entertaining and interesting travelling companion and would tell me stories of his early days as we drove on our way to other estates.

One of the highlights of his shooting programme was the Royal weekend at Luton Hoo, the very lovely estate owned by Sir Harold and Lady Zia Wernher. Lord Louis was very particular about time, and quite often our departure time would be set at five minutes to three or five minutes past three; very seldom would it be three o'clock or quarter past or half past. He liked to plan to arrive at Luton Hoo some fifteen minutes before the Royal party to enable him to be in the welcoming gathering, so I would be outside the front door of Broadlands with the car – filled with petrol, the luggage, guns and cartridges and the dog – all ready to be off, five minutes before the scheduled departure time. His Lordship would come out of the house twenty minutes later and pile into the car saying, "We're late," in such a manner as to give me the feeling that it was *I* who was late! Then it was foot down and fly. The speed limits were often ignored on the motorway and we usually managed to arrive at Luton before the Royals.

On the journey Lord Louis would read all the newspapers or record a speech for another function he was due to attend. He also had the happy knack of being able to put on eyeshades, recline the seat and fall asleep at once, and for a set time. He often asked me to wake him when we reached a certain point on the route, but I never had to as he roused himself at the appointed time. He also did breathing exercises which nearly caused me to have an accident the first time I was driving for him.

These exercises consisted of taking long, slow, deep breaths and noisily exhaling rapidly, at the same time bending forwards. We were late as usual, and were roaring up the motorway at high speed when Lord Louis slumped forwards with a loud gasping noise. It really frightened me and I swung the car across the road to the hard shoulder, braking violently. His Lordship suddenly sat up, demanding, "What the devil's going on?"

"I thought you were ill," I said.

"Nonsense," he replied, "Just doing my exercises." After that I was prepared for the grunts and groans coming from the passenger seat, but never quite got used to them.

The house at Luton Hoo is very large and only one half of the building was used by the Wernhers as living quarters. The remainder of the rooms was given over to "The Collection" as it was known. This was in fact one of the finest collections of Fabergé in the world and was open to view by the public at certain times of the year. When we arrived, the butler, Bill Donohue, dressed in black jacket and pinstripe trousers, met us and took Lord Louis and his luggage, while I was shown downstairs to the staff dining room.

There I met the head gamekeeper, who helped me carry the guns and cartridges to the gun room. The guns were taken from their cases and assembled, put in a rack and locked up securely until the following day when the shooting began. I was next taken to the staff lift that whisked me to one of the upper floors, then down a long corridor to a room with a label on the door proclaiming that the occupier was "Lord Mountbatten's Man". The room was huge, containing two large beds, vast wardrobes, a hand-basin and dressing table. The view from the two big windows was superb, overlooking the gardens at Luton Hoo with their magnificent topiary and flower beds.

When we sat down for our dinner in the evening it was like taking part in an episode from the TV series, *Upstairs, Downstairs*. The butler, formally dressed, sat at the head of the table, and we sat in order of precedence around it, honoured guests above the salt and the home staff below. We were served with great formality with no Christian names being used. After the meal we adjourned to the more relaxed staff lounge and chatted until bedtime.

Shooting the following day was excellent and beautifully organised. The birds flew well and were shot in good numbers, making a most enjoyable day. Once, on our return to the house, Bill Donohue gave me the privilege of showing me the dining room set for the Royal dinner. It was truly magnificent. A huge crystal chandelier dominated the room and lit the table, set with glittering silver. Candelabra, also silver, were placed centrally, containing smoke-grey candles. I was told that Lady Zia did not like white candles as they reminded her of funerals, so she had them dyed grey. The splendour of the room, and indeed of the house, impressed me greatly. It was like being transported back in time fifty years. I consider myself very privileged to have seen and taken part in a life-style that is fast disappearing. Soon after our last weekend at Luton Hoo, Lady Zia died and now the big house is not used. Alas, it is the passing of a splendid age.

On our way home to Broadlands, we would sometimes stop to visit Mrs Barbara Cartland, who was a great friend of Lord Mountbatten. She is a tremendous character, very kind and thoughtful. As we were leaving after my first-ever visit to her home, she shook hands as we said good-bye, thanked me for looking after Lord Louis and gave me two of her books, autographed with a note of dedication to my wife. Those books are kept as a part of the many happy memories of my travels with Lord Mountbatten.

We would then journey on and stop for tea with another friend of His Lordship's, Sir Robert Neville, a Royal Marine General. Lord Louis regarded the Marines with high esteem, only one down from his much-loved Royal Navy which is praise indeed.

Driving back through London was always a bit of a traumatic experience for me, being very much aware of the passenger I was carrying, but he was an excellent navigator and usually we traversed the city with no trouble. On one occasion the problem was technical rather than navigational. Rain poured down on inky black roads, and we had just come into the outskirts of London when all the electrics in the car failed. I coasted the car to the kerb and luckily found myself opposite a garage. I ran across the road and found the proprietor, a charming elderly Indian who ran the garage with his young son. I explained our predicament, at the same time letting him know that the President of the Royal Automobile Club was my passenger. Donning waterproofs, they came over to the car, very soon diagnosed the trouble and in a few minutes more put it right. They then required a signature for the RAC form, which Lord Louis duly signed. He then rolled down the window and shook hands with the grimy hands of the garage men. The elder of the two was quite overcome and, with his eyes full of tears, told his son to remember this day for the rest of his life, for he had shaken hands with the last Viceroy of India. It was spontaneous actions such as this that endeared Lord Louis to people from all echelons of society: he always dealt with them at their level as equals.

The Broadlands shoot continued to improve until on one day more than two thousand birds were killed. The day caused a lot of controversy in the shooting world. Words like "disgusting" and "disgraceful" were bandied about, but is it? Without shooting and the need to rear pheasants, there is no doubt that there would be very few, if any, of those lovely birds left in our country.

There has always been a poaching problem with the pheasant, and they are fairly easy game to poach. The traditional picture of the poacher as a dear old countryman, taking a few pheasant or fish for the pot or beer money, no longer applies. We did have one old-fashioned poacher in Romsey in the 'fifties, known as Yorkie. He was a great character, usually dressed in knee-breeches, leather gaiters, tweed jacket and a wide-brimmed hat. He carried a knobbled stick and his well-trained dog always ran with

him. More often than not it was the dog that was seen first as Yorkie just blended into the countryside. We would spot him working a piece of cover and give chase. His favourite trick was to jump over a barbed-wire fence onto the railway line, which to him and us was no man's land, because that area belonged to British Railways and we had no power to arrest him once he was there. After some discussion with Lord Mountbatten, he contacted the railway authorities and we were granted permission to go onto their property for the purpose of apprehending poachers. Warrant cards were issued to us and we waited for Yorkie. We finally had our day and he was arrested on the railway property. He was very upset and thought it was a rotten trick for us to produce warrant cards, so destroying his safety zone. I think it really broke his heart, as we never saw him again on Broadlands land and he did in fact give up poaching. He spent the remainder of his life directing traffic in the town.

Long gone are the Yorkie-type gentlemen poachers. Today there are highly organised gangs that travel great distances in the pursuit of game. They kill large numbers of birds when they hit an estate and are doing it for sheer profit. They also stop at nothing. One of our keepers out on patrol approached some men at three o'clock one morning. They were obviously up to no good at that time of night. As he shone his torch on the group, one of them put up a gun and fired down the beam of light. The keeper was severely wounded in the head and was very fortunate not to be killed, so it can be seen that the modern-day poacher deserves no sympathy at all and should be dealt with in the harshest way the law allows.

Although I had had no experience of handling shotguns before coming to Broadlands, it was necessary for me to be able to use one to control vermin on the river. Accordingly, I went to a gunsmith in Salisbury to see what he had to offer, cheapness being my main concern. I was rather tentatively asked if I would mind having a hammer gun. There was no objection on my part, for at that time (1956) gunmakers were getting into production after the war had ended and hammerless ejectors were coming onto the market. These were very much in demand, so buyers were trading in their old hammer guns in favour of the modern ones. I was shown three guns, all with hammers, and chose the cheapest and in my view the nicest, for the price of £27. On my return home I examined my purchase more closely. It was a beauty, with fine, brown, Damascus barrels and engraving on the lock. I was very pleased then and could not wait to show it to Lord Louis. He suggested I should bring it with me next time I accompanied him on a clay pigeon shoot. This I did and His Lordship had a try with my gun. After a few shots he pronounced it was a good one and that he could shoot better with the old hammer gun than with his very splendid Purdey. I did offer to swap, but after a little thought he declined!

That old gun had a rough life in many ways. I kept it in our main fishing hut, when one night the hut was broken into and my precious 12-bore was stolen. It was found three days later, hidden under a hedge a mile from where it had been taken. The exposure to the elements for the time it had lain undiscovered had done the barrels a certain amount of harm, but once I had cleaned off the surface rust and given them a good dose of oil, they looked a lot better. The best thing was that the gun still worked well.

One of our fishermen, who was also a gunsmith, spotted my old gun in the corner of the hut where I kept it and asked if he might have a look at it. He then jokingly offered me £120 for the gun, saying it was in fact an early Purdey and I should find out more about it. I telephoned Purdey's in London and quoted the number stamped on the gun. "By Jove," the man said, "that's an old one." Having looked at their very comprehensive records, he told me the gun was made in 1881 and, if in good condition, should be quite valuable. I have since had it thoroughly overhauled and put in proof. Now I have it locked away, for it is ever-increasing in value – probably the best investment of £27 I shall ever make.

CHAPTER SEVEN

THE SALMON CLOSE SEASON is the time when we tackle the river maintenance, although I realise that most people think river-keepers hibernate for this period. Pile-driving the banks to stop erosion, building groynes, breakwaters, bridges and fishing platforms and shifting the large quantities of silt that have accumulated during the season: we do all this work ourselves and very rarely have to call in outside contractors. Some of the jobs we carry out are massive tasks to the uninitiated. This is where I was fortunate in working alongside Len and Sid Pragnell, who were brothers employed on the river staff for many years. They were splendid men of the country with a great knowledge of doing things in the manner of years gone by before there were tractors or many mechanical aids to help with the heavy jobs. Len was tall and well-built and assumed the role of leader, while Sid was shorter and stockier. Both had a delightful sense of humour and were dedicated to the river.

We had no motor-mowers for bank-trimming at that time. It was all done with the scythe, the sort of tool carried by Old Father Time. There is a great skill in using this cumbersome, awkward-looking blade, but Len and Sid were masters. I must admit that when I looked down the river watching them at work, they nearly always seemed to be stopped. What they were doing in fact was giving the blade a rub with the stone to keep it razor-sharp. The trimming they got through in a day would do credit to a modern mowing machine, except they would do a better-finished job, with all the mowings taken back from the path. They gave of their knowledge freely and, with great patience and a lot of laughs, taught me the art of scything. Most importantly, they showed me how to adjust the movable hand grips for my particular height and reach. In this way the scythe became "mine".

At the end of a day's work in the summer, Len and Sid invited me to go "quodding" with them. It was explained that we would be going out at night, quodding for eels. I was a bit suspicious at first and thought it was one of their leg-pulls on a townie. But it was no joke, although it certainly provided many laughs. The quod consisted of long garden lobworms, threaded through with worsted string and rolled into a ball about nine inches in diameter. A long bamboo pole was used as a rod, obtained from our local furniture shop having been used for rolling carpets. From the end of the pole a length of stout string was tied and then, with the ball of worms fixed to it and a heavy weight suspended about a foot below the worms, our quod was ready for use.

Sid's method was to drop the worms into a weed-free patch of water close to the bank and lower it until just clear of the bottom. After a little while the eels could be felt attacking the ball of worms, and with a steady progressive heave, the quod would be lifted from the water and onto the bank. There would be several eels attached to the quod, their teeth having stuck in the worsted strings. Not being able to untangle themselves quickly enough, they were out in the field before they knew what was happening. Len and I had the job of finding the eels and putting them in a wet sack. Trying to gather up live, wriggling and very slimy eels in the dark is a difficult job and to hold them long enough to put them in a sack I found almost impossible until Len showed me the way to pick them up with one hand, using the middle finger as a clamp between the fingers either side. It is amazing the difference this makes in handling the fish: they just cannot slip away.

Sometimes we would quod from the punt and on one splendid occasion, after a very successful couple of hours' work, we had nearly a sackful of eels. Having got to the bank, Sid heaved the sack onto his shoulder and the bottom fell out of it, filling the boat with a great wriggling mass. In our attempts to fill a new sack we slipped and fell amongst the fish, helpless with laughter and covered in eel slime – a memorable night indeed. There was one time when a very large and active wasps' nest in the river bank caused a bit of a hazard for fishermen and Len, Sid and I were despatched by Walter Geary to destroy it. The method to be used was to pour a spoonful of Cymag down and around the hole at the entrance to the nest. We arrived at the site, keeping a very safe distance while the Cymag powder was carefully put onto the spoon. The next step was to be rather tricky as it meant going very close to the nest to pour the powder into the hole while many wasps were still flying in and out of their nest. Naturally we were all reluctant to approach too close, but finally Len took the spoon and, stooping low, crept forward. He poured the powder into the nest and retreated rapidly. When we looked round, Sid was about a hundred yards away. We pulled his leg for being cowardly. "Arr," he said. "I don't like bees, and I hate hornets, but I hates wapsies even more."

In my first close season we had need to replace Webb's Bridge. The bridge consisted of a single baulk of Canadian pine wood, nearly sixty feet long and fourteen inches square, weighing about four tons. It was supported on each side and the middle by means of three large piles driven into the river bed and a crosshead of ten inches by eight inches. Had I been asked to put such a timber across the river I think I would have hired the biggest crane I could find for the task. With Len and Sid's know-how the job was done by the three of us in just under the hour.

The preparatory work did take some time. Three sets of supports had to be completed first. This entailed rigging our pile-driver, known as "the Monk". It was a large structure about twenty feet high, made of wood with two vertical rails for the three-hundredweight lump of solid iron used as the driver to run down. The weight was hooked onto a wire and hauled to the top by means of a winch, then a pull on the release rope and the great piece of iron slid down the runners and thumped the pile into the river bed. Because of the very size of the Monk it took nearly five days to erect, so putting in the supports of the bridge was quite a long job.

The mighty baulk was delivered and dropped from the timber carriage close to the river. The next task was to put it across and lay it snugly on the supports. I had no idea even how to attempt it, but for Len and Sid there was no problem. Using long oak levers and blocks of wood, the baulk was lifted from the ground and six-inch-diameter wooden rollers slipped under. Slowly the baulk was rolled over the river until its point of balance was reached. The end projected from the bank at an angle of about thirty degrees upstream. Next, our two large working punts were lashed together side by side and a trestle made across them that was two feet higher than the bridge supports. The baulk was levered over until it dropped onto the trestle, the punts then floated across the river and the baulk was in place. I know it sounds a simple thing to do, but without my friends the Pragnell brothers, it would certainly have been a structural engineering problem.

That piece of Canadian pine, incidentally, cost £120 in 1957 and Lord Mountbatten would bring his grandchildren to view, as he put it, the most expensive bit of wood in the world. I think His Lordship would have had a fit had he known the cost of a similar baulk in 1982. For after twenty-five years Webb's Bridge needed replacing again. This time I was unable to buy Canadian pine – 55-foot-long timbers are quite difficult to find nowadays. However, I finally located one of suitable dimensions. It was not pine but greenheart, a very dense hard wood weighing nearly six tons and, once dried out, very nearly impossible to drive a nail into. Speed was essential so we were able to cut corners by rolling the new bridge over the old one, jacking it clear and then cutting the old bridge away with chain-saws, enabling the new bridge to be dropped in its place.

The cost of just that one piece of wood was £1,300. And fishermen complain when their rents are increased.

The Monk has now been retired. It was a dangerous piece of mechanism, as I know to my cost having caught two of my fingers in the unguarded cogs of the winch and mangled them badly. I replaced it with a tool that was originally made for putting in fencing posts. Tractor-mounted onto the hydraulic system, it takes the form of a large hammer with a 200-lb pyramid-shape weight on the end of a fifteen-foot beam. We had to weld on strengthening bars, but it is now a very efficient mobile part of our river equipment enabling us to drive far more piles in one day than in a week with the old Monk.

At the top end of our water in a lovely old mill house named Sadlers Mill. There is a certain amount of controversy regarding the spelling of Sadlers Mill. It is mostly written as "Saddlers" today, but I found in my old game book a printed bill-head and it is spelt "Sadlers" with only one "d". It is dated 24 October 1914 and is a note to the Broadlands agent from the mill lessee saying, "I have received your rent demand and I cannot possibly settle it at present with so many outstanding debts and trade very bad. So trust Mr Ashley's kindness for a later settlement." The poor old boy must have been having a hard time financially.

Alongside the house is a series of hatches used to control the height above and the flow through the mill. Many years ago the old water-wheel had been removed and in its place a water-driven turbine was installed. This supplied electricity to much of the Estate. In more recent years the turbine has been removed and replaced by two power-driven hatches, electrically controlled to open and close automatically as the river rises and falls, so maintaining a constant head of water upstream. Below the hatches there is a large pool where the salmon gather, awaiting the ideal conditions which urge them to make the run up, through the rushing water pouring through the hatches, to the shallow, quieter water higher up river where they will spawn.

The usual time salmon choose to make their run is after heavy rain in late October. It is a wonderful sight, the fish leaping high in the air and throwing themselves forward. They quite often used to damage themselves on the brickwork surrounding the hatch tunnels until we padded them. This important job has always been known to us as "putting the bags up". The bags are in fact sacks filled with straw which are nailed around the archways. It was a three-man job: Len, Sid and me. Sid was "up-top man", who was responsible for holding the punt close to the wall and for keeping an eye on the hatches to prevent them being inadvertently opened while we were working in the tunnels. We did have a nasty moment several years ago when a hatch was accidentally pulled and the punt with Len and me in it rapidly filled with water. Luckily we only got very wet, but we now keep a very close eye on the hatch controls.

We quite enjoy "bag day" as we often have an audience asking lots of questions, which gives us the chance for a bit of leg-pulling. Like the dear old lady who asked why we were putting up bags of straw and when told, "For the salmon to roost in", gave Sid a shilling for being so kind and thoughtful. Nowadays our audiences are not so gullible.

Many people are still of the opinion that the salmon are leaping in frustration because they are unable to get through the hatches and I am often accused of actually preventing the fish from running. This just is not so. The fish will and can run through, but not until they are good and ready and not a moment sooner. About twelve years ago the river was very low after a hot, dry summer and there were about two hundred fish lying in the mill pool waiting to negotiate the hatches. These fish were perfectly visible in the gin-clear water and they were making no attempt to go through. I became more and more concerned about such a large number of salmon being on view and expected poachers to arrive at any moment to take advantage of these easy pickings. I thought we might encourage the fish to move up by creating an artificial spate. This was done by closing all the hatches to build up the upstream water-level to almost flood height, then fully opening one hatch. The result was most spectacular! A jet of water poured through, colouring the river below and giving the appearance of a full spate.

I was most impressed, but not so the salmon, for after a couple of hours when the water had cleared they were still in the pool in serried ranks with only a few having gone up. I tried the same idea the following day but it did not fool the salmon one bit; in fact the situation was aggravated, as my artificial spate seemed to have drawn even more fish from below into the mill pool.

By this time it was mid-October and still no rain. My anxiety was such that I considered asking the river board for permission to net the fish and lift them over the hatches to speed them on their way. But as so often happens, Mother Nature looked after her own, and saved me the mammoth task of transporting all those fish, by raining for thirty-six hours non-stop, resulting in the river rising by several inches. It wasn't a very big rise of water but it was fresh and there was enough of it to stir the fish from their lethargy, and away they went. In two days the pool was clear, with only the odd fish or two showing. Over the years I have come to the conclusion that it is the fresh water of a spate and not just sheer volume of water that decides when the fish should make their spawning run.

At that time there was built into one of the hatch tunnels a wooden construction described as a salmon ladder. This fish-pass consisted of three narrow dark chambers, each separated by baffle-boards. The idea was that the salmon should swim into the first chamber where he could rest awhile, then swim under the baffle-board into the next chamber, rest again, and so on until he reached the river above. In theory this was fine, but I think the whole thing was too narrow and the chambers too small, which created a great deal of turbulence and the wrong conditions for salmon to rest. The efficiency of the fish-pass was in such doubt that a photo-electric cell fish-counter was installed to count the number of fish passing through, in the hope that it would establish the merit or otherwise of the salmon ladder. One morning the recorder showed that an incredible number of fish had run up during the night and that they were in fact still running at the rate of approximately one every two minutes. This called for closer examination. When the covers were removed from the centre chamber there was not a sign of all those salmon running; instead, there was an empty quart cider bottle going round and round in the turbulent water. On every revolution it passed the electric eye, and so recorded another salmon going through. This only proved that the photo-electric eye was working 100% efficiently; I am afraid the same cannot be said for the fish-pass.

Eventually the wooden sides of the chambers began to rot and looked dangerous, so the whole structure was removed. The fact that it has gone does not seem to have slowed the salmon's progress up river; on the contrary, the fish seem to be arriving at the higher reaches earlier than ever before.

Sadlers Mill is the last of the more formidable obstacles which the salmon must negotiate before having a comparatively free run to the spawning redds. Yet it was only twenty or so years ago that, on the higher reaches of the Test, the salmon was regarded as vermin to be destroyed at all costs. There were dreadful stories of keepers upstream shooting salmon with 12-bores or wiring them out with rabbit snares; even pitchforks were used to destroy their enemy. This war against the salmon took place on the jealously guarded trout waters. In fact, about forty years ago there was a very strong move by the owners of these trout waters to put a blockade at Sadlers Mill to prevent the salmon from running above Romsey, because they believed that salmon ruined their trout fishing by disturbing the trout eggs when cutting their redds. This may well have been the case as salmon spawn later than trout, and in those days the fisheries depended on the natural reproduction of the trout to stock their waters. But nowadays these same fisheries rely almost entirely on trout hatched and reared to takeable size in their very efficient hatcheries and fish farms. Therefore the keepers and owners look upon the salmon with a less jaundiced eye and allow them to spawn in peace.

Autumn visitors to Romsey, after looking around the beautiful old Abbey, may take a very pleasant walk to the Memorial Park and, by crossing two delightful side-streams, reach the salmon leap at Sadlers Mill. It is quite a tourist attraction and indeed, on some fine weekends in October, it is almost impossible to drive a car along the Causeway for the hundreds of spectators gathered there, many of whom come from as far afield as London to watch and in the hope of photographing the salmon jumping. One chap, a keen amateur photographer, repeatedly telephoned me, asking, "What time are the salmon going to jump?" I tried to explain that the fish do not work to a timetable, but he still persisted in telephoning. Eventually I became so exasperated that I just invented a time and told him the salmon would be jumping that day at 3.25 p.m. This was obviously the information he required and he rang off a very satisfied man. Later that evening he phoned again in order to thank me and to marvel at the accuracy of my forecast which had enabled him to take some splendid photographs!

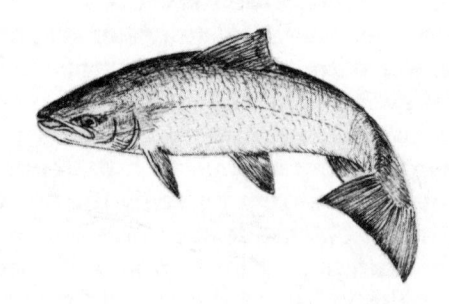

Chapter Eight

In 1964 the much-feared salmon disease became apparent in the rivers of Ireland. Many fish died from it. I remember seeing horrific photographs, with hundreds of fungus-covered salmon drifting downstream and rotting carcases on the banks. We all prayed that the plague would never come to English waters. At the time the disease was named "Columnaris" after the fungus that grew in columns from the open lesions on the fishes' bodies. Later the scientists investigating the outbreak renamed it Ulcerative Dermal Necrosis, or UDN for short.

We saw no sign of it at Broadlands but I had been concerned at the then declining numbers of salmon being caught by our fishermen. From 1956 until 1964 our average catch had been 350 each year. Then in 1965, 1966 and 1967 the average dropped dramatically to 190. The prophets of doom forecast that the end of salmon fishing was in sight, and in my view it certainly seemed possible. Something had to be done quickly, for should the salmon cease to run the Test, Broadlands would have miles of fishless water and I would be out of a job. I explained my worries to Lord Mountbatten and proposed that we should start a trout rearing scheme so that in the event of the salmon fishing fading out, we could stock our river with trout, and convert our fishery into a very attractive trout water. He agreed, and I went away to prepare plans for a mini trout farm.

The original plans I drew up were in fact for a very large fish farm, even by today's standards, but would have been too expensive to build for a new venture. It was proposed to make a small one consisting of three long ponds, and to grow on fish that we were to buy in as fry. The site for the ponds was chosen carefully. It was an ideal position between two of the old water meadow carrier streams, which ran parallel about eighty yards apart. One stream came directly off the main river and the other formed a land drain. The main river carrier eventually curved round and crossed the other stream by means of an aqueduct built more than a hundred years ago as a part of the water meadow system. The best feature of this site is that in the eighty yards separating the two carriers there is a difference of eight feet in water levels, always ensuring a good flow of water.

My original idea was to have earth-sided ponds, but I had to alter this as soon as digging began. The machine had excavated a trench six feet deep and twelve feet wide in solid peat. The operations of the digger shook the whole area and it was necessary for the sides of the trench to be supported. Our problems quickly multiplied. When we arrived on site the next day the section of trench that had been dug was full of oily-looking water. A gaseous smell, which I suspected was methane, emanated from the peat. We stopped the water seepage by building a crude coffer dam at the top end of the pond, then carried on with some messy digging until the first pond was roughly in shape. While the machines were on site they dug out second and third trenches in readiness for stage two of the scheme.

After shaping the sides and bottom of the first pond by hand, I decided to use corrugated iron fixed to wooden posts rather than concrete, which would have been more costly. The first pond was in two sections divided by a concrete wall with drop-boards to maintain the water level in the upper section, and screened with wire mesh to prevent the fish escaping. The inlet and outlet were also built of concrete, with mesh screens and drop-boards to control the flow and level. The whole pond was surrounded with a half-inch-mesh chicken-wire fence, the bottom of which was buried twelve inches deep to prevent predators from burrowing underneath it.

With the major building work completed, the problem was what to do about the bottom of the pool, a black, muddy peat exuding a gaseous, oily mixture. I covered the base with six inches of chalk, well rolled down, followed by three inches of three-quarter-inch washed gravel. Finally poles were laid across the ponds on the top of the fence posts and wires set nine inches apart stretched longitudinally to prevent the herons and cormorants from destroying the new fish on arrival.

At last my first "stew pond" was ready. Each of the two sections could hold 500 fish up to 2½ lbs each. I decided that for the first year I would purchase fish of about four to six inches. As I had to buy 1,000 fish, I planned to try 500 brown trout and an equal number of rainbows. It was a decision I had cause to bless many times in the months that followed. I ordered the fish from Somerset for delivery on 2 May.

On 28 April all was ready. I dug away the last barrier of peat and slowly began to flood the pond. The water flowed through and cleaned the residue of peat and rubbish from the sides. I was very happy and satisfied when I went home that night, but on arrival next day, I saw that the water had worked under the concrete outlet of the pond, softening the peat it stood on, and the structure was tipping up. Once the full weight of water got under it, the whole thing would slide gently into the low-level carrier.

With 1,000 fish arriving in four days' time, we had to work quickly. The first job was to shut off the water and drain the pond, thus relieving pressure on the outlet. The entire concrete structure then had to be levered back into its original position and a raft of concrete pumped underneath it to make a larger area of support. We went home that night weary, and hoping that everything would be all right when we let the water in again. I eased out the first drop-board on 1 May and allowed the water to run in slowly until the level had built up enough to allow full flow. The keeper's "wee angel" breathed on me: the outlet structure held – and has done to date.

The first batch of fish – 500 brownies – arrived on 2 May and I put them in the lower section of the pond. They were fine fish, gleaming golden-brown, from the water of the Exe. They were followed by 500 rainbows, splendid fish, which were put into the top section. The fish seemed content and took a feed on arrival. This was surprising because brown trout often will not feed for a couple of days after a long journey, whereas the rainbows would feed in the transporter tank.

At last the work was finished and I was satisfied with the result. Stage 1 was a reality.

On the morning after stocking the stew, I discovered two fish dead on the screen. Two more were dead by lunchtime and a further four the following morning. I examined the dead fish externally and internally, but could find no obvious cause of death. I spent a number of hours observing the live fish, but their behaviour seemed quite normal. Just after feeding time I noticed that several fish were acting strangely, almost as if they were drunk. I fed them again watching them very carefully and found the answer to the mortalities. It was simple really, though not something that would present an obvious hazard when building a stew pond. This is what had been happening: the food given to the fish consisted of pellets, about the size of a small pea. These pellets float, making the trout rise to take their food and at the same time getting them used to surface feeding. There was a very good flow of water going through the ponds, about 5,000 gallons an hour, and when the pellets were thrown into the ponds some were swept down each side. The little fish taking the food with such eagerness at the sides of the pond were in fact stunning themselves on the iron corrugations. I cured this problem by reducing the flow of water through the stews and by feeding the pellets down the centre of the ponds. Most of the stunned fish had recovered but a few had drowned, and these accounted for the mystery deaths.

Trout kept in stew ponds always seem to feed voraciously even though there is enough food for them all. The main reason seems to be that when there are about 500 trout in a small area, the competitive element for food is very strong indeed and the fish sometimes feed even when they are not hungry. The remarkable thing is that once they are put into a river or lake and have to fend for themselves, they become after a few days almost as nervous and wily as a wild fish.

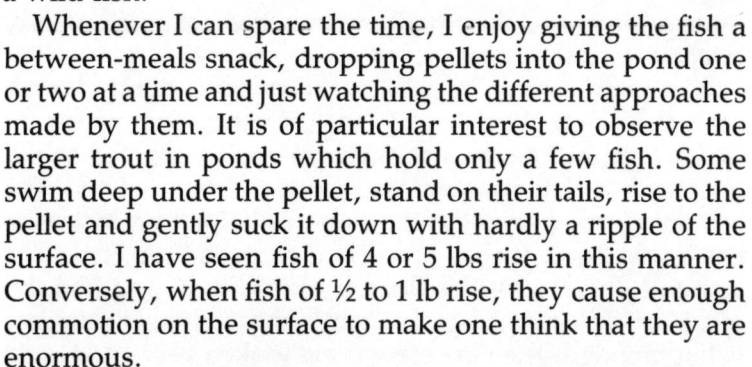

Whenever I can spare the time, I enjoy giving the fish a between-meals snack, dropping pellets into the pond one or two at a time and just watching the different approaches made by them. It is of particular interest to observe the larger trout in ponds which hold only a few fish. Some swim deep under the pellet, stand on their tails, rise to the pellet and gently suck it down with hardly a ripple of the surface. I have seen fish of 4 or 5 lbs rise in this manner. Conversely, when fish of ½ to 1 lb rise, they cause enough commotion on the surface to make one think that they are enormous.

Once we had overcome the problem of the fish dying after feeds, things went smoothly and the trout grew rapidly. I then had to consider the building of my next stew. At this stage I must stress that the whole trout rearing scheme was to be built as economically as possible and by using our own labour force of three men. With the experience gained in the construction of the first stew in which the corrugated iron sides had proved unsatisfactory, an alternative had to be found, a material strong enough to prevent the peat sides eroding.

After much searching, a Southampton plastics firm came up with the answer – butyl sheet. Butyl is a black plastic material, not very thick but immensely strong; if torn, it can have a patch welded on in situ. All in all it was the ideal material, and reasonably priced, too. The manufacturers came out to the site and measured the stew. The sheet was to be all in one piece to cover the sides and bottom for the full length of the pond. We had to prevent the sides being gnawed through by rats so, before the butyl sheet was laid in position, the sides were covered with three-quarter-inch-mesh chicken-wire, making life very difficult for rodents.

The manufacturers came to lay the sheet in position, as I thought it best for the professionals to do that part of the job. They arrived with the sheet all folded concertina-wise and, starting at the bottom of the stew pond, unfolded it as they came towards the top end. The method used for securing the longitudinal edges of the sheet was simply to dig trenches two feet from the sides of the pond, lay the butyl edge in and then refill the trenches – simple but very effective, for it is guaranteed for ten years.

This pond was to be a two-section one, slightly longer than the original stew. All the concreting for the centre divider, with the screens and drop-boards, as well as the inlet and outlet, was laid on top of the butyl sheet. The bottom of the pond was the same as before: six inches of chalk with three inches of washed gravel on top. This was also put down with the sheet underneath it, and had the effect of tightening up the butyl sheet along the sides. When flooded, the new stew pond looked very good, and this time we had none of the problems of end-pieces washing away. One of the side benefits of using a plastic lining on this pond is that weed growth does not adhere so much to it; also, the chalk deposit from the water, which has to be scraped from the corrugated sides of our first pond, can be scrubbed from the butyl sheet quite easily with deck scrubbers.

By November 1968 the fish in the first stew pond had reached 1½ to 2 lbs, and very fine-looking fish they were. As the water temperature dropped, the feeds were cut down to one a day, and the fish began to fight as spawning time approached. The fighting is at times very vicious, and it is common to see fish with great raw wounds or with flaps of torn skin trailing behind them. The problem with these injuries is that Saprolegnia fungus grows on them and, if allowed to thrive, will completely cover the infected fish and eventually kill it. The standard treatment for fungus is a very strong anti-fungal dye called "malachite green". A short dip in a strong solution is usually enough to treat a mild infection. In some ways, any form of disease is worse in fish farms than in natural conditions, chiefly because in a stew pond the fish are overcrowded by comparison to fish living under natural conditions in the river. On the other hand, the fish in a stew pond are under the watchful eyes of the fish farmer. At the first sign of trouble, treatment of the disease can be started and quite often a cure effected, unless there is no known cure for the disease.

Late in 1968 I went to give the trout their morning feed and noticed that one or two fish were hanging about in the slacker water at the bottom end of the pond. On closer examination I saw reddish spots on the fishes' heads. I netted out a couple of these fish and had an even closer look. The lesions were soft and, when pressed, a watery pus oozed from them. There was also the beginnings of Saprolegnia fungus attached to the wounds. I first suspected furunculosis, a very nasty disease to have in a crowded pond, but at least one that could be treated and cured.

The following morning there were ten fish dead with similar lesions on the head and body to those on the fish I had removed the previous day. There were also a number of other fish in the pond showing the same symptoms. Only the brown trout were affected; the rainbows in the section above the browns were perfectly clean and fit and still feeding well. Over that weekend the condition of the brown trout deteriorated considerably, with a number of fish dead. On Monday morning the river board were informed and a doctor from the Ministry of Agriculture, Fisheries and Food came to examine the fish. The diagnosis was ulcerative dermal necrosis, UDN, the dreaded salmon disease which had already been confirmed on the river upstream of me.

The disease seems most virulent in low water temperatures, so it appears at its worse in the winter. All that I could do was to carry out the advice of the Ministry man and kill all the browns and incinerate them. It was a heart-breaking job, killing 500 beautiful fish of 2 lbs but, having completed it, I drained down the pond. I allowed it to dry out completely and limed the bottom and sides thoroughly. The pond was left like this for a month to ensure complete disinfection before being flushed out for a week and then re-flooded for use again. While all this had been happening the rainbows continued to grow and remain healthy, but of course I had only half the stock with which I had started.

With the great perversity of nature, the salmon run that year was excellent, the best for many years. So I had 500 very good rainbow trout and nowhere to put them. If I transferred them into the main river with the salmon fishing going on, they would very quickly be taken on salmon flies and baits. What should I do? Sell the fish live for stocking another river or lake? Such a notion did not appeal to me, for by this time I regarded them as my fish and had become quite attached to them. I liked the idea of a lake, though, and I cast my eye around the Estate for a suitable site of about four or five acres. Eventually I found a good place on a U-bend of the river where the lake could be flooded from an inlet of river water. The site was ten acres, but I reckoned that by using some of the spoil to build an island or two and a few promontories, I would finish up with a lake of about six acres. One big disadvantage of the site was that the lake could not be drained other than by pumping out: to be able to drain a lake or pond is a most desirable feature. It would be an enormous task, a bit daunting, but being used by now to trouble and strife and to overcoming problems, I believed we could tackle it. At length, however, owing to the high cost of the operation, the idea was dropped and I was left with only one alternative – a river or stream.

Once the decision to build our own trout stream had been taken, I had to find a suitable site with that most important asset – an abundant supply of good, clean water. After looking at several ditches on the Estate, I chose the low-level carrier which took the outfall from the trout stews. This was an ideal situation, since I would have complete control of the water flow.

The stream – or ditch, as it was then – was in fact a land drain with no inlet from the main river. The water already in it was a deep rust colour, with a heavy concentration of iron. When I consulted old maps and a plan of the old water meadows, I found that this particular ditch had been known as the "Red Water Cut", clearly named after the colour of the water.

I walked down the length of the ditch and found it a little daunting. The first mile was almost straight, and the second only marginally better. The whole extent was densely overgrown with reeds, and one section was only three feet wide with an occasional pool about ten feet wide. It looked at first sight a relatively big undertaking. Clearing the trees and other undergrowth would take a long time. Widening and deepening would have to be considered, which in turn would mean disposing of the spoil removed during the digging process. Before anything could begin, however, I had to negotiate with the farm management, for besides the machinery running back and forth during the various operations, I would need to make considerable inroads on the farm's arable land over a two-mile stretch. I had the good fortune to be dealing with a most co-operative farm manager who not only agreed to my having some of his land but also suggested that the spoil could be spread over the surrounding land, to be ploughed in at a later date. His suggestion reduced expenses considerably, for the alternative would have been to cart all the spoil by tractor and trailer and dump it a couple of miles away.

Now that the first two problems had been resolved, I had to consider the actual widening and deepening. A local plant-hire firm sent a mechanical digger on the agreed date, and I explained to the driver how I wanted the stream shaped into pools and bays, with some places left narrow to give the water concentration. These narrows would serve to increase the speed of the water flow and would liven up the stream a bit. I walked the length of the stream, accompanied by the machine driver, who then set to work and really used his imagination. He dug a pool round a clump of trees at one point, creating an attractive little island.

The digging was completed after a week, and the time came to increase the flow of water to give the stream a good flush. I allowed the water to run for a several days, and while this was in progress we cleared the banks where there were too many small trees such as alders and elders, and some holly bushes. With the spoil spread evenly by the bulldozers over the surrounding fields, and the banks trimmed and cleared, the ditch took on the semblance of a stream.

Because of the slope of the bed of the stream over the two-mile length, the water level would have to be built up a good deal, especially at the top end to give a reasonable depth for fishing. To create weirs for this purpose, I cut some stout boards from old railway sleepers and laid them across the first culvert, about half a mile from the top of the stream. I put in three twelve-inch boards, which built up the water level by eighteen inches about two hundred yards upstream. Between the first culvert and the top end I made a wooden dam to build up the level in the top end of the stream. This, however, was not a great success, as the water pressure was such that it washed under the dam and made a large hole in the river bed. I removed it and replaced it with one made of stone, which worked well and looked better. All subsequent dams were built this way, with either brick or stone rubble, and blended well with the surroundings.

When the height of the water had been raised to the required level, the river bed itself looked very bare. The stream needed fly and shrimp life to maintain a head of trout, and it was also necessary to provide the fish with cover. Since the stream had a good flow, I transplanted some roots of the ranunculus family, and long-leaved water crowfoot, from the main river. I used pegs, bent to horseshoe shape, to keep the roots in place while they were establishing themselves. Where the water was too deep for the pegs to be pushed into the bed, I simply tied the roots to a large stone and dropped them into deeper holes. In the same way I planted starworts, watercress and other varieties of pond weed along the sides. I raised the water level again, slowly this time, for I did not want to wash out my newly planted "garden", and in about a month most of the weed had taken root and was growing well.

Stocking posed a few problems. Although the stew ponds actually emptied into the trout stream, it was not feasible just to open the hatches and let fish swim out into the stream as they would have taken a long time to spread themselves out over the full length of the river. So, to maintain an even supply of fish in all parts of the stream, we would have to transport them and put in a few at a time down the whole length of the trout stream. I had a tank which fitted in the back of a pick-up truck, but as it was a comparatively small tank it meant that we were able to transport only fifty fish at a time.

When moving that amount of fish crammed into a small area, the oxygen in the water supply is very quickly used up and equally quickly, the fish expire. To keep the fish alive for the duration of the transfer from stew pond to river, the water had to be oxygenated. I used a cylinder of oxygen with a diffuser. This consisted of a brass pipe, sealed at one end, with holes punched into it at regular intervals along its length, and completely encased in a ceramic material much like the stuff from which the old red flower pots were made. When the cylinder was connected to the open end of the pipe, oxygen was forced through the ceramic covering under pressure from the cylinder, and into the water by means of the millions of bubbles that were produced.

One may wonder why it is necessary to go to all the trouble of having a diffuser. Why not simply put the pipe straight into the water? If this were done, very large bubbles would be produced; they would rise to the surface, burst and release all the oxygen into the air so that none of the life-giving gas would be absorbed into the water for the fish.

On other occasions I have used "breathing air" for moving fish on short journeys. This is the type used by sub-aqua divers, and has the advantage not only of aerating the water but of being considerably cheaper than oxygen. The latest method of fish transportation is the use of an anaesthetic which puts the fish into a state of suspended animation. The anaesthetised fish lie in the tank as though dead; as a result, far less damage is done to them than if they were swimming about and knocking themselves against the sides of the container. There is also no need for oxygen to be pumped into the container, as fish use less oxygen when they are unconscious. In this way they may be moved over relatively long distances. Once returned to fresh water the fish quickly recover and swim away no worse for the experience.

There were many applications to fish the new water and, being the first to realise that this small stream was not everyone's idea of trout-fishing, I asked each of the applicants to study the water before deciding whether or not to rent a season's rod. A few of the anglers who came to have a look thought it would be "easy fishing" but this turned out not to be so. The fish have to be stalked and will spook easily because, owing to the narrowness of the stream and its high banks, even the smallest fisherman must look like the Empire State Building and be visible a long way off. And yet there are days when one can almost jump on the heads of the fish and nothing seems to scare them, but at these times it is almost impossible to catch them anyway.

We opened on 1 May, and fish were caught. There were some good hatches of fly, and in the second week of May a fine hatch of mayfly. How or why they appeared on a new stream I do not know, but every year since then we have had mayfly hatches.

My next step in the trout scheme was to rear fish at an earlier stage, which meant buying in three-inch fry. Since the existing stews were too large for the convenient rearing of fry, we had to build a much smaller pond. We decided to make a fry-box twelve feet long, six feet wide and eighteen inches deep, longitudinally divided into three sections. This was made of one-inch-thick tongued and grooved boards, and sealed with a waterproof, lead-free paint. Baffle-boards were put into each section twelve inches down from the water inlet, to avoid too much turbulence in the water, and screens were incorporated in the baffles to prevent any rubbish from entering with the water. For the water supply we had to sink the box below the water level of the carrier stream, as we had to rely on gravity feed in the absence of electricity on the site for a pump. We dug a hole large enough to take the box with a bit to spare to enable us to walk round. The hole was three feet back from the bank and deep enough to take the box below water level. Having finished the excavations, we drove three-inch steel pipes through the front of the pit into the high-level carrier and, once the pipes had been cleared with rods, we had our water supply.

The fry-box was lowered into the pit, and a platform round it enabled us to walk in comfort. The fry progressed well, but we had a silt problem. I tried several methods of filtration, but with only limited success. The silt did not harm the fry, but it simply fouled the bottom of the box and created unnecessary work. The ideal water supply would, of course, have been a spring from a borehole, but it would have been a very expensive operation to drill down to the depth required to find pure water.

As the rainbows in the stews continued to grow, we carried on with the task of going through all the fish in a pond and sorting them into various sizes. I was surprised by the difference in growth rate in a batch of 500 fish, and it was necessary to grade them. Had we left them all together in one pond we would have ended up with a few enormous fish, many of medium size and many more little ones which had hardly grown at all. By separating them as soon as it was apparent which fish were growing quickly, the small fish would be given an opportunity to feed and develop so that we would eventually have a number of good-sized fish of about 1½ to 2 lbs. They tend to be free-rising, and I think they fight better. They are easier to rear from a fish farmer's point of view and most of all, they are cheaper to buy for stocking. Yet all the time the demand seems to be for bigger and even bigger fish. These are all right in their way, but they are expensive to rear, and the bigger the fish the more trouble they are to keep in the stews during the winter months as they seem more prone to the various diseases that appear in the colder water. If fishermen wish to catch large rainbows the fish farmer can supply them, but the angler will have to pay for them.

Ultimately, I wanted to strip fish and rear them from the egg stage, but in the meantime I experimented with hatching fish in the fry-boxes. I made a couple of trays to fit into one of the fry-box sections. They measured twelve inches by eighteen inches by one and a half inches, with the sides made out of bullnose architrave timber and the bottoms of very fine, green plastic netting of the kind used for fly-screens. I then cadged a few hundred eggs from a keeper upstream who had his own hatchery. The eggs were spread evenly over the trays and then placed in the fry-box with a gentle flow of water over them. A few eggs died during the first week, but most of the others did well. The water temperature at that time fluctuated sharply, prolonging the incubation period. Then to cap it all, torrential rain fell for a week and the ensuing floods brought down masses of silt which completely smothered the eggs. By the time we had cleaned them up about eighty-five per cent were dead. Some of the remainder eventually hatched and the alevins thrived, but they were much further behind in their development than those which my colleague hatched in spring water. There was no doubt about it – it was imperative for me to have a spring, even if I drilled for it myself. And that's precisely what I did.

The drilling of a bore-hole by a professional firm would have been costly, so I read up as much as I could in the local library about the subject. It did not seem a difficult task to undertake. The question was where to bore: it was not simply a matter of drilling where I wanted the water to be. How was I to discover the whereabouts of a spring? The obvious answer was a water-diviner.

I have always been sceptical about water-divining. I simply could not see how a chap wandering about with a twig in his hand would be able to feel water at a depth of thirty feet or so underground. But I discovered that a local farming firm employed a water-diviner, and he was reputed to be very good. I telephoned him and he agreed to come out and see if he could find a spring for me.

A red mini drew up and a tall, impressive-looking man got out. He had a look at the site, went to his car and produced four straight twigs, each about twelve inches long. He took one of the twigs, bent it at right angles and, holding this in his hand down at his side, he walked along slowly in a straight line. He stopped suddenly and thrust the twig into the ground. He then turned sharp right, took another twig and set off again in exactly the same way. After about ten yards he stopped again and, as before, thrust the twig into the ground. Sharp right again, the same measured pace and, in turn, the third and fourth twigs were "planted", so forming a square. At this point he returned to his car for a fifth and longer twig which he formed into a U-shape. Taking an end in each hand, he walked in a zig-zag within the square he had made with his original four twigs. Suddenly he halted, and the U-shaped twig vibrated and turned upwards. "There's your spring," he said, "and not very deep – about thirty feet."

He suggested that I try my hand at it, which I did, only to discover that I did not have the power of water-divining.

Before we could begin boring it was necessary for us to obtain permission from the river authority to extract water from a three-inch borehole. One is legally bound to obtain the river authority's permission, and a good thing too, otherwise everyone would be drilling holes in the ground and taking water before it reached our rivers.

The water-diviner suggested that we use a three-inch pipe to bore down on the place he had indicated, and we would need six five-foot lengths. The first section was the most important because it would be the drill as well as the water-finder and outlet. To make the drilling-head, the end of the pipe had to be formed into a point and welded to give a good seal. Slots, six inches long by three-quarters of an inch wide, were cut longitudinally at regular intervals along the length of the first five-foot section of pipe. These slots were to allow the water in – if we found the spring. The top end of the section was threaded, and the remaining sections of pipe were threaded at both ends and joined together with collars. A solid steel head was made, which screwed into the top of each section to prevent any damage to the thread at that end during the driving process. It was essential that during driving the driving-head was screwed up very tightly to avoid damaging the threads.

At length we started drilling. The water-diviner stressed that the pipe had to go down vertically. Each section of pipe was put down, and after three days we were down thirty feet. We connected a diesel pump to the end of the pipe protruding from the ground. After pumping for about half an hour there was no sign of water. I phoned the water-diviner for advice. He thought that perhaps we had gone down too far, and suggested that we pull up a few feet.

"How the devil am I to pull thirty feet of three-inch pipe up out of solid peat and clay?" I asked.

"Quite simple," he said. "Just jack it up."

I borrowed a big clamp from him, which fitted onto the pipe with a bar across it at right angles. He also lent me two ten-ton-lift screw-jacks. These were placed on planks of wood under the bar, and we screwed them slowly upwards. We were to run the pump after each foot of pipe had been pulled up.

After two feet had been recovered the pump was connected again and a trickle of muddy water appeared. We drew up the pipe another foot, reconnected the pump, started it and there was my spring – water a-plenty, very muddy, but water. After an hour's pumping the water began to clear, and after two hours it was flowing crystal-clear. I was delighted. I put a maximum–minimum thermometer into the water, and the temperature remained a constant 51°F. To ensure that the flow would be maintained we rigged a supplementary diesel fuel tank on the pump to enable it to run all night. We pumped for five days and nights non-stop, with no reduction in the flow. During this pumping I took three samples of the water and sent them off for analysis – one to the river authority, one to our public analyst and the third to a laboratory in London.

While awaiting the analysts' reports I turned my mind to the problem of raising water from the borehole. A diesel pump could not be kept running indefinitely, and an electric pump could not be used because of the lack of electricity on the site. So it was back a hundred years – to the old water-wheel. This could be rigged in the high-level carrier, and connected to a reciprocating pump by means of a reduction gear. Winter would be a dangerous time, for if there were icy winds the water would freeze on the wheel paddles; the build-up of ice could unbalance the wheel and eventually stop it working. However, this problem could have been solved by building a simple shelter and lining it with straw.

These thoughts about water-wheels and pumps were premature, as I had not yet received any reports on the tests of the spring water. When at length they arrived I was surprised to learn that the fine, gin-clear water was virtually devoid of oxygen and full of iron salts. The lack of oxygen could easily have been remedied by spraying the water into the stew over a grating, so breaking it into tiny drops and oxygenating it. The iron salts were an altogether different matter.

I went to see one of the analysts, who was most helpful: "The water is quite useless for rearing fish, although it is drinkable – in fact, beneficial to human beings. The best thing to do with it is to put it in fancy bottles and sell it as spa water at five bob a bottle." This may well have been sound advice, but I had no wish to start a hydro or health farm. I just wanted my spring. I tried an experiment to precipitate and filter the iron salts; but it proved impractical because of the large amount of salts present. The reason for them was that my site was entirely on peat and, as the water flowed through, it picked up the iron salts from the peat. The only way to obtain pure water was to drill down into the chalk. Unfortunately the chalk stratum was something like three hundred feet deep, although three or four miles upstream it was only forty feet down.

The only answer was to devise a method of filtering the silt from the river water and trying to hatch eggs again. I dug a pit which I filled with fine gravel and allowed the water to trickle through slowly. This scheme worked quite well for a while, but the amount of suspended silt in the river was such that the filter bed soon became blocked and ineffective. Even today, I'm still looking for an efficient method of removing the large amounts of silt from the water.

Chapter Nine

As THE YEARS went by, the pressure on our river was ever-increasing from all angles. The demand for fishing became greater, so it was felt that we had to try and meet that requirement. Instead of resting the river on Fridays, which had been the practice for many years, that day was let also, meaning that we now fished seven days a week. We thought this would have an improving effect on our catches as well as giving more people the opportunity to fish, but there was certainly no increase in numbers of salmon caught. In the past fishermen sought to fish Saturdays, thinking they would derive benefit from the rested Friday, but it seemed to make little or no difference as no more fish were killed on Saturday than any other day of the week. With no rest day it did, of course mean that instead of thirty there were now forty people fishing every week. I don't honestly think this increase hurt the fishing a great deal, for rod-caught salmon make up a comparatively small percentage of the fish running a river and should not seriously reduce the stocks.

However, it must be a contributing factor, along with all the other pressure put upon the river system as a whole. There is no doubt in my mind that the character of the Test has changed quite dramatically in the last fifteen years or so. In the past the river would rise after a heavy rainfall, become coloured, then clear to a beautiful translucent green after a couple of days and retain its height for a week before very slowly dropping. Now, after a good storm the river may rise as much as eighteen inches in a few hours, turn a thick yellow in colour, drop nine inches overnight and within a few days without rain be back to what we used to term its summer level. I know that an eighteen-inch rise may sound very small compared to rivers like the Wye where they experience fifteen-foot rises, but for a chalk stream, eighteen inches means a great amount of extra water coming down. The causes of this change into a mini spate river are probably abstraction and so-called "improved" drainage.

In the old days, rain fell on the land and ran off into ditches and some of it found its way gradually into the river, but a lot of water soaked into the ground and slowly permeated many feet down into the aquafer, to bubble up months later in the form of springs which ran all summer, ensuring a nice steady flow and height to the river. These days, many of the fields are drained by means of a herring-bone system of pipes buried only three or four feet underground which carries the water very quickly to large ditches and thence to the river. This water, because it doesn't have to soak down very deeply into the earth, is comparatively unfiltered, so brings with it a fair amount of the fertilisers and other treatments that farmers spread over the land. Some of these "goodies" in the water, phosphates and nitrates in particular, ensure a very prolific weed growth in the main river which must influence the fly and shrimp life in the water considerably.

Then there is the amount of new building going on in the area. Instead of rain falling on open land, it falls on impermeable roofs and roads, flows directly into pipes and with great speed into the river. The combined use of these "improvements" means that the run-off of surface water is greater in volume, and so the river rises very rapidly.

Abstraction for industrial and domestic users also puts a great demand on the river system. Around twenty million gallons each day are taken by direct abstraction, but the method I find more worrying is the use of boreholes into the aquafer itself. This is a very insidious method of taking water before it even has a chance to become a part of our river, and in many ways I am sure the full effect is not really known. What does happen when water is continually pumped from the chalk hundreds of feet underground? Is it possible that some of the deep-running caverns containing water collapse and compress when the level is dropped, so making them unable to hold further water supplies? I don't know, but I do put it forward as a likely explanation of the decrease in flow of spring water.

Our river is now supplemented by the discharge of treated sewage. While agreeing that this treated effluent is clear water and capable of sustaining human life, it is not suitable for fish as it is almost devoid of oxygen and full of phosphates and nitrates which again act as fertiliser to weed; more often than not it is the wrong sort of weed growth like blanket-weed, which has a tendency to smother other good, life-holding weeds.

Now it has been discovered that the very rain that falls may contain a high proportion of industrial acid. How is this affecting fish and their propagation? It is probably too early to know, as work studying acid rain is still in its infancy, but it could be causing an imbalance in the sexes of fish born in the river. I know that in 1980 we caught 125 salmon with the idea of stripping them for artificial breeding. Of that number only four were male and, try as we might, we found no more cock fish. By just walking the river and seeing a lot of fish it was a fair assumption that it would have been a good spawning year. However, if the ratio of cocks to hens in the river was the same as we discovered in our catch-up, there would have been a great lack of fertilised eggs, so making a bad spawning season. If these conditions had prevailed for a year or two previously, a dramatic decline in the number of fish hatching would be the result.

I do wonder for how long we have deluded ourselves into believing that, because a lot of fish are seen in the river late in the year, a good breeding year will result. The very manner of salmon spawning is really pretty inefficient, which is probably why nature is so prolific in supplying many thousands of eggs per fish. Assuming that a pair of fish do spawn successfully and a fair percentage of eggs is fertilised, there are the dangers of scavenging eels and other coarse fish, more than prepared to vacuum up as many eggs as possible. The likelihood of extra-heavy water at that time of year may either wash away a redd or smother it with a deposit of silt.

The biggest disaster for the salmon was the discovery of their feeding grounds off Greenland. There the stocks have been plundered by fleets of fishing boats with their very efficient methods including sonar detectors and nets of incredible length. Thousands of tons of salmon of all sizes, many of them immature fish, have been taken. Most of this deep-sea fishing is carried out by countries with very little interest in the salmon, other than as a harvest to be reaped as speedily as possible, and with no thought of helping in the replenishment of stocks in the rivers. It would improve things considerably if a limit on their catches could be agreed with the possibility of a levy put on the numbers of fish landed. The money gained from the levies would help finance a series of salmon hatcheries, and river restocking could be programmed to ensure the continued existence of the Atlantic salmon as a species.

Unless something is done along these lines, and done quickly, then in my rather pessimistic view there is a very real danger of extinction. One has only to look at Europe of fifty-odd years ago, when there were about thirty rivers all carrying good stocks of salmon; today only four or five remain, and those aren't very prolific producers. I realise that in the case of many of the European rivers, the main cause of the salmon demise is the gross pollution of their waters. But as has been so magnificently achieved on the Thames, a great clean-up could and should be carried out before it is too late.

Many riparian owners of salmon fisheries, especially in

Scotland, do have a comprehensive restocking scheme and are showing good returns. While most of the evidence regarding the benefits of artificially rearing salmon for stocking is purely circumstantial, there is no doubt that the number of fish-running rivers where restocking has been practised over several years has greatly increased, as in the case of the Eriffe fishery in Ireland. Lord Brabourne began hatching and rearing salmon there in 1977, when the catches on about ten miles of river averaged seventy fish per season. By 1982 the year's total had risen to more than 400.

On the Test in the 1950s a few salmon were artificially reared to feeding fry by Sir Richard Fairey's keeper, Reg Dade. I met Reg in 1959 and found he was still interested in rearing salmon. We began catching up the adult salmon at Bossington, which is about sixteen miles by river from the estuary. Reg had a fine, spring-fed hatchery house where the eggs were put to hatch. Over a six-year period we averaged about 200,000 salmon fry each year put into the river. Now a coincidence it could be, but from a couple of years after we began our restocking, the runs of grilse in the Test grew in numbers, so much so that they made up the bulk of our catches. Then, sadly, Reg died and with him the enthusiastic help I needed to carry on with our stocking programme. The interesting point is that now, the grilse runs are diminishing alarmingly.

There are other factors that may also have an effect on salmon runs on our river: the work that goes on almost continuously in our estuary, dredging deeper channels for the ever-larger container-carrying ships, and the very building of the enormous container berths a few years ago which did in fact reduce the entrance to the Test considerably and altered the runs into the estuary.

Yet another pressure on the river system is the rapid proliferation of fish farms. By their very nature with the necessity for high and unnatural stocking densities in their ponds they cause to appear in the water a great deal of suspended solids and chemicals which would not normally be present in the river.

However, I do not think that the blame should be laid on

any one of these adverse happenings on our river: it is the cumulative effect of them all combining to change the ecology of the river and so make the fishes' environment less attractive to them. For if, as is the popular belief, salmon use their sense of smell to return to the river where they were born, it could be that due to changes in the make-up of the water, the scent does not go so far out into the salt water, making it impossible for the fish to smell their way home. These are purely theories, attempting to explain in some way why the salmon runs have declined in recent years.

One of the more recent major changes to our river environment is the building of the M27 motorway. There have been plans for a major new road since 1930, known in those days as the "Proposed Great South Coast Trunk Road". Having lived in the shadow of this threat of upheaval for so many years, everyone had thought it would never happen. But in 1974 it all started. Men came with their theodolites and began driving pegs along the river bank by our boundary where the M27 was to span the river. They were quickly followed by armies of men with their gigantic yellow earth-moving machines. It was a rather horrifying sight to see them appearing over the hills in the middle of the unspoilt countryside. They gave me the feeling of being in the presence of invaders from another planet.

A great gash was cloven through the hill and steadily the scar in our countryside approached the river. When it reached the bank, the first operation was to straighten the river and parallel the sides. Steel interlocking piles were driven in and tons of concrete poured behind them. A large Bailey bridge was thrown across the river for a working access. That, I thought, would be where the problems would come, as it was an ideal platform to fish from. Sure enough, it wasn't long before we began to find night lines suspended from the bridge. These I reeled up, then I snapped the hook in two and cut the line into very small lengths, leaving a heap of chopped-up nylon for the poachers to find in the morning. This developed into a bit of a game with the motorway workers. Although they were a big, rough, tough bunch of chaps, I must say we had very little trouble with them during their building operations; in fact, we developed a good and happy working relationship.

There was a certain amount of river disturbance but it seemed to have little effect on the fish life. I never cease to marvel at the amazing tolerance of salmon. Quite often when fishing, one has but to move clumsily or cause a rod to flash in the sunlight to scare every fish within many yards' range. But there, lying a little way out in the river, right opposite the pile-driver, were two salmon, side by side, seemingly oblivious to the shock waves that must have been transmitted through the water. The fish were also completely uncatchable! I think they were deaf, blind and nerveless.

The commotion did appear to upset other forms of country wildlife. Coots and moorhens moved upstream, apparently to get away from the noise. Their absence was only temporary, as they soon accepted the racket that was going on and slowly crept back downstream, but leaving a healthy no man's land between them and the construction work. One little animal which appeared to leave the area entirely was the mole. I imagine that the tremendous vibration and shock waves through the earth caused by the pile-driving must have upset the poor moles, for they packed their bags and moved north about a mile. One of our meadows took on the appearance of a ploughed field thanks to the great many mole-hills being dug by the migrants. Even now, some eight years on, very few of the little creatures have returned to the neighbourhood of the M27.

As the road building progressed on the western side of the river, it would very effectively cut off about eighty acres from the rest of the Estate. The motorway building contractors approached Lord Mountbatten for permission to extract gravel from eighty acres south of the road. As very little could be done on the land during the construction works, permission was granted and the diggers moved in. The usual practice at most workings, once all the gravel is extracted, is to backfill the enormous hole with topsoil and then plant a forest of small trees, which seem never to grow – probably due to the lack of drainage as all the gravel has gone. The M27 gravel extractions were only a few yards from the main river and three streams crossing the fields had to be diverted away from the workings, so it wasn't surprising that, soon after digging began, water started seeping into the hole. It was on seeing this water seepage that I had the idea of turning the finished hole into a fishing lake instead of filling it in as usual. I mentioned my idea to Lord Louis, who saw the potential of the lake and agreed that negotiations should start with the motorway builders. They were very happy and quick to give their consent. It was an ideal arrangement for them: they would save a lot of money by not having to cart all the topsoil back from the dump, and all the diggers were already on site. So we set about finalising the shape and size of the lake.

Bearing in mind that there were eighty acres of land to be worked on, we could have finished up with an area of water covering some fifty acres. This was not to be, for one morning the mechanical diggers were taking their enormous mouthfuls of soil and dumping them into lorries when a little man appeared and asked the digger drivers how much further they intended to excavate. On being told the next forty acres, he said, "You'll have to watch our gas main, then." It was a mercy he had appeared, because the contractors were completely unaware that a twenty-eight-inch gas main was buried there and ran across the full length of the proposed diggings. It was considered impossible to carry on excavating any nearer to the twenty-eight-inch pipe, so our proposed lake was reduced in size by half. We decided to have a couple of islands in the lake, one large and a smaller one. The inlets and outlets with their water controls were designed and plans were drawn. Before submitting them to the planning authority for permission to build, we had to decide what we would do with the completed lake.

I wanted to establish a large day-ticket coarse fishery because there was a great need for one in our area. His Lordship suggested a trout fishery, which I did not really want or think practical, as we did not have the facilities to rear enough trout to stock a large area adequately, though my main objection was the task of policing the water. The lake is situated on the boundary of the Estate and would have been wide open for poachers to come in and remove our trout. It would need an army of men to keep them at bay. I felt that poachers would not want to waste their time and energy going after coarse fish.

Every avenue was explored, even to a firm of "leisure consultants" being called in. They produced a splendid report on the amenities that could be crammed onto the site. These included fishing, sailing, model boating area, camp site, caravan park, a kiddies' farm, putting green, a museum and the pièce de resistance, a restaurant on stilts over the water. All very wonderful ideas, but we didn't really want a holiday camp, or to give it the name chosen by the chaps who formed the plan, a "leisure activity facility". We all had a very expensive laugh and submitted our plans for a coarse fishery.

We had quite a few problems with people the planning authority sent out to our lake site. The first couple of lads they despatched to view the area were rather "arty" and visualised "hills there, flat areas here". We reached some sort of agreement with them, but the next couple that appeared wanted the contours changed. I became rather angry, which seemed to set our planning permission back a few months. In the meantime a tree-planting programme had to be thought out and submitted. Again we had problems as the planning officials saw acacia trees and other, more exotic plants growing all over the place. Our head forester very soon put them right, and a much more natural selection of trees was arrived at, all varieties that grow like weeds in the Test Valley.

At long last the happy day arrived when the planning authority gave their approval and work was able to start. The lake was to be a coarse fishery only and no other use could be anticipated. No cars other than for the disabled or our own vehicles were to be allowed in the lake area, so a large car parking space was cleared outside the lake boundary. The construction went on apace and early in 1976, the finished lake was inspected and accepted by the Estate.

The sluices were opened and slowly my lake filled with water. I was a bit daunted, as twenty-seven acres of water looks a vast area, especially when it needs stocking with fish. At that time most coarse fish were hard to come by and quite costly, and we needed a great many fish. On the Estate we did have some old gravel workings which had flooded, forming several rather smelly, dirty ponds. Youngsters were allowed to fish them, because it kept the kids away from the main river and gave them somewhere to go and enjoy a little fishing. These boys would occasionally come to me with stories of big fish they had seen or caught. I put most of their reports down as true budding anglers' exaggeration. Now we were desperately seeking fish, I decided to examine the ponds more closely. They didn't look very promising. The pools were surrounded by banks of dead reeds which had grown in, reducing the area of water considerably. The water itself was filthy and covered in a fairly thick film of oil which had come from the many old oil-drums which had been thrown into the ponds. However, we decided to give it a go with our electric fishing gear.

Almost as soon as we began fishing along the reed-lined verge, there was a commotion in the water and to our astonishment a carp of about 4 lbs came to the surface. In a couple of hours' fishing we had twenty in our tanks. In the next few days we electric-fished the ponds five times, taking out carp and tench and transferring them to our new lake. The largest carp caught was 8 lbs. The mystery of how the fish had got into the ponds was solved later, when I heard that a year or two before, during the building of the new road, a lake had to be filled in. Luckily for us, the driver of one of the machines doing the work was a fisherman and, as the lake was reduced in size, he observed the fish crowding into ever shrinking water. The fish were caught up in nets, put into dustbins and the driver released them safely into the nearest patch of water that he knew of, our old gravel pits.

As the year of 1976 went on, it developed into a long, dry and very hot summer. It was fine for holidaymakers but very soon it became a disaster for everyone concerned with fish and fishing, and indeed water. In many ways we were fortunate with our lake because it was very deep in parts and it did have the benefit of three streams running into it. Even though the flows were very greatly reduced, the level of the lake was maintained. Many fisheries in our area were really suffering, especially the smaller lakes. With water temperatures soaring into the eighties, fish stocks were dying as the oxygen in the hot water diminished. Every day on our local radio I heard of lakes drying out and many dead fish.

I telephoned the local radio station and let it be known that we had transport, tanks and oxygen equipment and would be prepared to go anywhere to rescue fish in trouble and pay something for them as well. Our rescue service proved a success, as we certainly prevented a great many fish from perishing during that hot summer, and so the new Broadlands Lake came into being with a very good stock and a great variety of fish species.

Now the lake is a successful fishery and in my view will soon be one of the premier waters in the country. It has also become a haven for wildlife, especially birds. There is a great variety of duck coming in with some of the rarer divers and waders too, so we really have something to thank the motorway for: the price of progress is not always harmful to our environment.

In 1981, which was designated the Year of the Disabled, we were approached by an association for the handicapped to see if we could provide facilities enabling wheelchairs to come close to the lake for their occupiers to fish. After some consultation with wheelchair users, we built two large platforms, slightly raked towards the bank, with a nice level path leading to them. A road and car park were also made and the whole area reserved for handicapped fishermen. It was a lovely idea and it means that many people who were previously unable to can now enjoy fishing our lake.

That drought of 1976 also had a terrible effect on the river and the salmon fishing. The water level was desperately low and the flow-rate greatly reduced. We recorded temperatures that were previously thought impossible in a flowing river. The weed growth was phenomenal and river management became very nearly a guessing game. Should the weed be cut, so allowing the water level to drop even further, making the river unfishable? Or should the weed be left to grow, maintaining a little bit of water height but equally unfishable? These conditions prevailed for many weeks. The few fish that were in the river were all but uncatchable, being very lethargic and completely uninterested in any fly or bait presented to them. What damage to the fish and fly life that summer did is difficult to assess, for when the rain did come, it came in plenty, causing the river to rise rapidly to almost flood level, washing away the banks of weed and giving the poor old river a much needed scour-out. How many dead fish went to sea with the high water I do not know, but there must have been many bodies lodged in the tangle of weeds. We can only assume that possibly four generations of salmon would have been decimated during the drought and we may well still be suffering from that disastrous year.

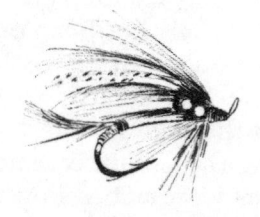

CHAPTER TEN

LORD MOUNTBATTEN did sometimes come out fishing, but he was not really a keen fisherman. However, he showed great interest in the river and in the people fishing the water. He would often stop whilst out riding to pass the time of day with our fishermen, and many of our clients have happy memories of these little chats. One day several years ago, I was talking to one man fishing on our Rookery beat when we saw a flash of light in the distance. It was as if someone was using a mirror as a heliograph. We couldn't make it out. Slowly the flashing got nearer to us and it turned out to be His Lordship on his horse. He was dressed in casual clothes, but on his head was the silver helmet worn by the Life Guards in Whitehall. He was Colonel-in-Chief of the regiment and as such took part in the annual Trooping the Colour ceremony. With his usual thoroughness it was his practice to get used gradually to wearing and riding in his uniform, as it wasn't exactly the normal clothing one would wear. The helmet, even without the splendid plumes, made vision very difficult. I asked him if he enjoyed the splendour of the occasion and his reply was, "I don't see much of it with this hat on!" He would ride round the Estate wearing first the helmet, then a few days later the plumes would be fixed on. Then, after a further period, he would add the very heavy breastplate and finally the long, stiff boots similar to thigh-waders. The whole uniform was cumbersome and very hot to wear, quite an ordeal for an elderly man, but he told me he would not have missed taking part in the Trooping for anything.

I had never seen Trooping the Colour and His Lordship, knowing this, gave me tickets for my wife and myself. We had excellent seats in the second row of the Admiralty reserved section. The weather was gorgeous and I must say I was really concerned for our Lord Louis sweltering in all his heavy uniform. But I needn't have worried as he sat there the whole time with Prince Charles and Prince Philip, completely unruffled. It was a truly magnificent spectacle, the sort of parade that we in Britain do so well and rather take for granted. I will remember it always, especially as it happened to be Lord Mountbatten's last Trooping.

In my earlier years at Broadlands, Lord Louis held the position of Supremo and one weekend a year was given over to the Chiefs of Staff. This was a working and fun weekend with many of the famous wartime service chiefs at Broadlands. We reserved two salmon beats for their use and much enjoyment was had by them. Sir Dermot Walsh, who was then Chief of the Air Staff, came out fishing. He was a big man and was wearing a long service mackintosh. It was a very windy day and cold with it. Standing on one of the fishing platforms, he made a mighty cast. He must have become slightly off-balance, for the wind blew under his coat and in a trice he disappeared into the river. He bobbed up at once, still clutching the rod, and we hauled him out. I will never forget seeing him being transported back to the house, spreadeagled on the back of a Volkswagen Beetle.

During my service with Lord Mountbatten, many famous people have stayed as his guests at Broadlands. Some of them fished and I have been fortunate to meet them and look after them while fishing. One day His Lordship phoned me to ask if any fish were about as he wanted to show the Royal children a salmon. The Royal children were Prince Charles and Princess Anne. They duly arrived and we all trooped off upstream to High Bridge, which was a rather rickety old cattle bridge. There was usually a very good salmon lying by the centre supports of the bridge and it was possible to see the fish quite easily in the crystal-clear water by lying flat on the bridge decking and spying through one of the many cracks in the floorboards. I looked down and sure enough, there was a fish of about 12 lbs tucked under the bridge, just waiting to be caught.

I had better explain here that over the years we had perfected a drill for removing salmon which decided to lie in this awkward position. This is how it was done. The fisherman would lower his bait from the upstream side of the bridge and I, the spotter, would look through the crack and, when I could see the bait, give the necessary instructions to go left or right a bit, or downstream, or deeper. It was a fascinating way to observe the fish's reaction to various baits and I don't know which was more exciting, being the fisherman or the spotter. The real problems started when the fish had taken the bait. It invariably turned and ran off downstream, leaving very little hope of the fisherman ever being able to bring the fish up. So I would pick up the line on the downstream side of the bridge with my gaff. The fisherman then slackened his line, putting two half-hitches round the reel handle, and dropped the rod into the river, and I would then rapidly pull it through under the bridge. As long as the two people concerned know what to do, the whole operation can be done in a matter of a few seconds, and while I do realise that giving a fish a slack line is dangerous practice, in the position we are in on the bridge, there is little alternative, and we have lost very few fish using this method.

I carefully explained all the moves to Lord Louis in the event of him hooking the fish, and emphasised that he should not just hang on to the fish but let it run downstream if it wanted to. I then showed the fish to the Royal children and all was ready, with three of us prostrate on the bridge peering through the cracks and His Lordship ready to lower the bait down to the fish, which seemed to get bigger the longer we looked at it! I had put on a nice fresh prawn for His Lordship to present to the fish as they are easier to see in the water for the spotter. As the prawn appeared in our view the excitement mounted. I guided it slowly nearer to the fish and, when the bait was about eighteen inches from his mouth, the fish suddenly moved forward. We saw his mouth open and the prawn disappeared into it. As the fish turned I bellowed, "Strike!" But there was no need. The fish was "on" and thrashing madly on the surface. "Let him run!" I shouted but I'm afraid the damage had been done and the hook had pulled out. There was great disappointment, and the ten-year-old Prince said, "*My* Daddy wouldn't have lost the fish."

That was my first meeting with Prince Charles. There have been many since on the Test at Broadlands. He has become an excellent fisherman, especially enjoying the fly.

We also had visits from foreign diplomats who sometimes showed an interest in fishing, although not necessarily knowing anything about it. One weekend I remember well, two parties came to the river. One group of gentlemen with their wives were from Spain, the other party hailed from Paris. In all they numbered a very valuable twelve. I took them to Black Dog, one of our best holding pools. As soon as we arrived, a salmon of about 15 lbs head-and-tailed off the lower breakwater. An absolute cert, I thought. There was only one man in the whole group who wanted to fish; the rest were photographing or spectating. The would-be fisherman among them asked me to have a try for the fish. I hooked it on my second cast and, amidst much shouting and laughing from the spectators, played and gaffed a fine fish of 16 lbs. The whole episode had been filmed, and I have since seen the film and a good record it is.

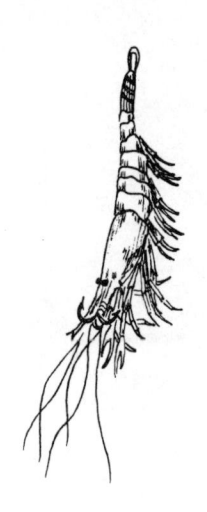

As I removed the fish from the river, another, slightly smaller one rolled over in midstream. The member of the party who wanted to fish came forward and told me he wanted to try, and was it possible that he might even hook the fish? I said, "Of course you'll catch the fish," with a deal of confidence I didn't really feel.

"If I catch him, a bottle of pink champagne for you," he said.

Fine, I thought, without much hope of ever seeing it. I showed him how to use the reel and, after three or four tries, he hurled the bait in about the right place. With the usual perversity of fish and fishing, his bait was taken at once. The noise from our audience was quite deafening and the poor chap holding the rod couldn't understand my instructions so the fish was given too much stick and totally mishandled. Finally, everything having been done wrong, the fish was gaffed! When I took the hook from the fish's mouth I discovered that it had taken the bait so viciously that it could have stayed on for a week and would never have come unstuck. The whole party was very happy, none more so than the man who had caught the fish.

About four months later, Lord Louis telephoned me, asking that I should go to the house. When I arrived at his study he presented me with a magnum of pink champagne from the delighted French diplomat. I kept that bottle for nearly a year and we drank it at a family party celebrating our tenth wedding anniversary.

It is well known that Lord Mountbatten was a very forward-looking man; not only did he have a very inventive mind but he was uncannily accurate at seeing the potential of any idea put to him. He was also mostly proved right in his assessments, which probably didn't endear him to some of the deadheads of power in Whitehall. Our country owes much to him in very many ways, some of them generally unknown. So in a very small way I became involved with one of the age's most revolutionary developments in transport. I had received a message from Lord Louis summoning me to the house for an informal meeting with him and, as was his way, he came straight to the point. Although our talk was without formality, it was, he stressed, to be completely confidential. He then related the following story.

He had met a man in the corridors of Whitehall who had shown His Lordship an invention which, in Lord Louis's view, had a great future in both service and civil use. The inventor had apparently been trying to get the authorities interested in his new mode of transport for a long, long time with no result, and was becoming angrier and more frustrated daily. The poor man could have readily sold his idea to a foreign power, but was determined that Britain should be the developers and reap the benefit of his invention. My part in the scheme was to fix him up with some fishing to give him a relaxing outlet for his energies while Lord Louis was selling the idea to Whitehall. It was arranged that I should meet the inventor at the fishing hut and he duly arrived in an old red Jensen sports car. He introduced himself as Christopher Cockerell.

He has fished with us since that day, and in that time his hovercraft have come into being, speeding ferry services all around the world and enabling cargoes to be transported over terrain previously inaccessible to all other forms of vehicle. Although the inventor of this wonderful form of transport has since received a knighthood, it is little enough reward for his persistence and patriotism in keeping the hovercraft for Britain. We are fortunate as a nation that Lord Louis met Sir Christopher in Whitehall all those years ago.

In 1977 Lord Mountbatten announced to us that he was considering opening Broadlands House to the public. It came as a bit of a shock to me and the more I thought about it, the more horrified I became. We had several Estate meetings on the subject where many of my fears were allayed. I learned that it was not proposed to have sea lions and hippos swimming in my river or boat trips to see monkeys and gorillas on the islands, so the house and grounds opening became more acceptable, even though I was to lose about a hundred yards of fishing off the lawns immediately in front of the house. I was still concerned, and could imagine hordes of people out of control rampaging up and down the river with children falling in all over the place and tons of rubbish being thrown into the water. I needn't have worried at all: with Mountbatten thoroughness the whole operation was planned and organised to a degree where disruption to the normal working of the Estate is kept to the minimum.

The length of river involved was fenced off to prevent people falling into the water, and apart from that the opening to the public has had no effect on my department, although I am still taken aback when I go up onto the house beat and on rounding the corner see hundreds of people on the lawns. Prince Charles officially opened the house in May 1979, a very happy occasion on a beautiful sunny day. Lord Louis asked His Royal Highness for his pound entrance fee, which he did not have, so he borrowed the pound from Lord Mountbatten. Now the public have a new museum and many other attractions, all tastefully and well presented.

In August 1979 Marie and I packed our old Volkswagen Caravanette and set off to France for our summer holiday. We have an unwritten law that while on vacation we neither buy newspapers nor listen to radio news, so we are usually blissfully unaware of any events during our weeks away from home. Our journeys are mostly off the beaten tracks away from "touristy" areas, trying to find camp sites that are used by the French people. We find we can get along quite well with the language and meet some interesting folk. We had stopped for a few days at the charming little town of Limogne and had struck up a friendship with a French family camped next door. They spoke no English but we managed to converse well enough. The husband was most helpful in telling us where to go and see the local places of interest, especially the chateaux to visit. He was an ardent chateau visitor and I asked him if he ever holidayed in England. "Maybe one day," was the reply.

We suggested that when he came he should visit us in Romsey and we would show him my employer's "chateau".

He asked who was my employer and, on being told that it was Lord Mountbatten, he looked at his wife and said, "Ill est mort" – he is dead.

Thinking he meant Louis XV, like in French chateaux, I said, "Oh no, he is still alive and in residence."

"Non, non, terroristes!" he cried.

We rushed to the radio for news and heard that Lord

Mountbatten and three others were murdered and at least three more had been badly wounded. Stunned and horrified, we packed our van and set off for home at once. I drove through the day and into the night, boarded the ferry and crossed the Channel, arriving home at three o'clock in the morning. We had a certain amount of trouble getting into the Estate through the intense security that had been thrown around the perimeter. Once at home we learned of the enormity of the tragedy. Our dear Lord Louis, his grandson, Nicholas, Doreen, Lady Brabourne and a lad crewman were all killed in the explosion, with Lord and Lady Brabourne and their son critically hurt. It was all too much to take in. I just could not believe that the most black-hearted madman could have conceived such a diabolical plan. At Broadlands and in Romsey town it seemed that everything had stopped. There was almost complete silence. Even the birds were quiet, giving the whole place an eerie feeling.

Lord Mountbatten's body came home to Broadlands and thence to lie in Romsey Abbey, where the Estate staff stood on vigil. Everyone took a watch, even the girls from the offices. It was quite an ordeal. Thousands of people from all walks of life filed slowly past his bier, non-stop throughout the night. We watched the State funeral on television. The most deeply moving moment for me was when a close-up was shown of Dolly, his Trooping the Colour horse, with the long boots reversed in the stirrups. It was that picture which made me realise that my much loved and respected Lord Louis was no more.

We attended the interment at Romsey Abbey, and once again the whole town was utterly silent with a heavy feeling of sadness. The people of Broadlands were still in a state of stunned shock because of the murders and our hearts went out to the Brabourne family who had suffered most grievously. We felt a great sympathy for Norton, Lord Mountbatten's grandson, who was to inherit Broadlands in such a terribly sudden and sad manner. With his parents desperately ill in hospital and unable to help or advise him, he had to assume control of the Estate. He had already announced his engagement to Penelope Eastwood and the wedding date fixed for October 20. Bravely, and I think very wisely, they decided not to cancel the wedding and all arrangements would go ahead as planned. In the following months the Estate began to come to life again. Yet there was a feeling of a lack of warmth and a missing presence, although I don't think we were ever aware of these feelings before. On his grandfather's death Norton Knatchbull had become Lord Romsey and, having known him as "Norton" since he was a very young boy, I found it difficult to adjust to using his title and to saying "My Lord" when addressing him.

Marie and I, having returned from France so rapidly, had missed two weeks of our holiday. One of my fishermen, Alistair Sampson, kindly invited us up to Scotland for three days. He was there on a fishing holiday after salmon on the beautiful river Beauly. We gratefully accepted his offer and, having packed rods, flies, nets, etc., and complete with dog, set off for Inverness. We crossed the Beauly at Lovat Bridge and the water looked perfect, with fish showing nearly all the time. I couldn't wait to get at them.

The following day I was taken to fish the Falls Beat and on the ghillie's advice put on a tiny Stoat's Tail tube-fly. I waded into the river and cast out. A fish showed below me, then another opposite where I was standing. It was a long pool and as I looked down it, salmon were leaping all the time. I found great difficulty in disciplining myself to fish the run slowly, without hurrying down to the next fish that was showing. However, I did carefully cover the whole pool – without a single take. I couldn't understand it. I must have put my fly over fifty fish with no offer.

I sat on the bank and had a smoke, turning over in my mind what tactics I should employ. Having tried the fly that "always took fish" according to the ghillie, I thought, "What would I do on my own water?" The answer was to go bigger. The ghillie thought not, and advised me to go through the pool with same fly again. But I had already lost confidence in the Stoat's Tail and had a look in my hat for a larger fly. There I found a size 3 Thunder and Lightning, one of my pet flies. I tied it on under the disapproving gaze of the ghillie, and set off fishing again. My third cast produced a good strong pull and my first Beauly fish was well and truly on. It was duly netted by the ghillie and, to his amazement, was followed by two more. I finished my three days' holiday with eight salmon, all killed on my large Thunder and Lightning.

It was a delightful three days' holiday and we returned to Broadlands on 17 October considerably rested and relaxed, all ready for Lord Romsey's wedding which took place on 20 October 1979. The press named it as wedding of the year. It was certainly a glittering affair, with Prince Charles as best man. Wonderfully, Lord and Lady Brabourne were able to be there, though in wheelchairs. It was a lovely, happy day, although for many who were at the Abbey for the wedding there was a sadness too, for it had only been a very short while since we had all gathered there for Lord Mountbatten's funeral. Without a doubt, though, I could feel he was there, enjoying the occasion of his grandson's wedding to the lovely Miss Eastwood.

The Estate began to settle down under the new ownership and I felt it was rather nice to have a "Ladyship" back in Broadlands again after so long. We all were wondering what, if any, changes would occur in the running of the Estate now we had a young master in control. By and large things still went on as usual, yet I still expected to hear Lord Mountbatten's voice whenever I went into the house.

Our new Lord and Lady Romsey are great friends of the Prince and Princess of Wales. The boys had virtually grown up together under the watchful eye of Lord Louis. When the prince announced his engagement to Lady Diana Spencer as she was then, the wedding to take place on 29 July 1981, Marie and I were delighted to be honoured with an invitation to the ceremony in St Paul's. We were very excited at the prospect of attending such an occasion and our friend, Alistair Sampson, came up with an offer we were happy to accept. He invited us to stay with them in London and he would drive us to St Paul's Cathedral in his beautiful Bentley on the light-hearted condition that he could display and retain the special car park sticker we had been issued with, as his personal "status symbol".

We arrived at Alistair's house, had tea and then we set

off for Hyde Park to watch the pre-wedding day fireworks display. The crowds were enormous, but extremely good-humoured. We had a super picnic and a grandstand view of the really wonderful firework display. When it was all over we were literally swept with the crowd out of the park and eventually found our way back to the cars, but there was no hope of moving. So we had another mini-picnic until the traffic thinned out enough for us to drive home at two in the morning.

We rose at seven, feeling none the worse for our late night, or rather early morning. Alistair's wife, Camilla, produced a magnificent breakfast and, to start our memorable day, "Buck's Fizz" – champagne and orange juice. At last the time came for us to move off and, dressed in our finery, we climbed into the back of the shining Bentley. With our "chauffeur", Alistair, and Camilla in front, we felt very regal. I must say that I had a nagging worry about security, but I needn't have concerned myself, because as we joined the Royal route, the atmosphere generated from the thousands of happy people lining the roads I can only describe as love and happiness. It was really terrific. We were seated in the South Transept gallery, right opposite that lovely lady, Kiri Te Kanawa, who sang "Let the Bright Seraphim" so beautifully. It was a wonderful occasion and one we will never forget. When the ceremony was over it was back to Broadlands, where the Royal newly-weds were to spend the first three days of their honeymoon.

The Estate staff had been well prepared for their visit and the security was very strict. I had been warned by the police that I would be a target for pressmen because of the known keenness of Prince Charles for fishing and the possibility of him showing his bride the river that he sometimes fished. I had been issued with a special pass and a police radio for the duration of their three-day stay.

We arrived home from London at 2.00 a.m. Our phone began ringing at 6.45 a.m. It was, of course, the press. We had formulated a stock answer to any enquiries regarding the Royal honeymooners, which was: "They are on a private visit to Broadlands and as such will remain private." This reply did not satisfy the newspaper men, and my phone continued ringing all day. Various ploys were tried: one chap said he was phoning from Broadlands House and asked me if I was ready to go fishing with the Prince. We were prepared for this type of deceit and all I had to do was phone back to the house to learn that it was a false call.

It was the photographers that gave us our biggest problem, especially those from other countries. We had been warned that an attempt would be made by one group to enter the Estate by means of a canoe, paddling down the river. Then they would lie in wait in the hope of obtaining some informal photographs. Forewarned is forearmed, and we foiled their plan by stretching two wires across the river which were visible from the public road bridge, so they didn't even try the canoe. At 10.30 p.m. I did get further information from a friend, who had overheard a conversation in French, that the same group would be trying to break through the security ring using the river, but this time in frogmen's kit. I immediately radioed through the information to Police Headquarters, who in turn warned all the security officers on patrol. One of these officers then asked over the radio if any description was available of the men involved. There was a long silence and then a very laconic voice said, "No, but I would think that anyone wearing a frogman's outfit and flippers at this time of night would be rather suspect. Over and out."

But the attempt created a new problem as the frogmen intruders would have been very difficult to detect and, once on the Estate, there were many places for them to hide. Again, the "great water bailiff in the sky" must have been on my side, for as I was driving through Romsey I spotted their car and glimpsed a flipper sticking out from behind a wall. I called the police on my radio who very soon put a stop to their little game.

They still didn't give up trying. I was followed everywhere I went, the fishing hut was under constant surveillance, and we began to feel hunted. They came on motor-cycles over the fields, crawled along hedges, crept along ditches and culverts, climbed over walls. Their tenacity was really quite impressive, but the fact remains that not one of them got through. The efficiency of the police was absolutely one hundred per cent and how they kept their good humour I will never know, for after two days of being hounded, I had most certainly lost mine. I have a great deal of sympathy for anyone in public life who is constantly under similar stress. I only had a short time of it and was heartily fed up.

But it didn't finish there. The newspapers next day carried reports quoting me, and one even said I had been interviewed by their reporter, which was completely untrue. And then my daughter phoned me from Australia to say I had made the front page of the *Melbourne Times*. She read out the most sickly, sweetly-worded article I have ever heard. Imagine my feelings when I heard that "Bernard Aldridge [wrongly spelt] had lovingly taken Lord Mountbatten's rods from their velvet-lined, mahogany cases, the rods used by the Queen on her honeymoon"! I was then supposed to have "happily" called the Prince of Wales to tell him the day was perfect for fishing and the salmon were biting well. The article went on in this honeyed tone, and with each word my stomach went into tighter knots. The whole report was complete fabrication, but most likely believed by the majority of the paper's readers. I was chiefly concerned that the Family would hear of these reports and think that I had indeed given the interviews and "information" that had been published. I was so upset that I phoned Lord Romsey in Kent, but I need not have worried, as they know me well enough to realise that the way the article had been written was not my words and anyway they understood that the so-called inside information was rubbish.

So it was with a certain amount of relief we all went up to the house to witness the departure of the Royal couple for the next part of their honeymoon. It was a memorable, if hectic, three days for us all at Broadlands, and we were very glad to see the backs of the hundreds of media people that had plagued the Estate.

CHAPTER ELEVEN

I HAVE OFTEN been asked questions on my attitude towards the conservation of wildlife and indeed to the conservationists. I am, along with most river-keepers and gamekeepers, deeply involved in the conservation of our countryside and rivers. It is our life's work and of course, our very livelihood depends on maintaining the fish and birds needed for the continuance of our various sports. But it has to be done correctly.

When I have spoken with various groups of people calling themselves conservationists, they continually mention "the balance of nature", and insist that it should be allowed to carry on undisturbed. This is fine in theory, but nowadays there is no such thing as "the balance of nature" and there hasn't been for many years, due to the interference of man in a great variety of ways. I feel very strongly that now we must try to restore a natural balance, which is subtly different from leaving it all to Mother Nature. I do not wish to see my river devoid of so-called vermin or pests. We need pike, eels and other coarse fish. They help keep the water clear of weak or sickly fish and scavenge on dead material. It is the same with herons, cormorants and otters: they all have a part to play. What I don't want is hundreds of them. That, I think, is one of the dangers of giving a particular species complete protection for, given such immunity, in some areas they can breed to an extent where they become a pest. Prime examples here in Hampshire are the heron and swans: we have too many of them.

Conservation is getting rather out of control, as did the annihilation of animal life years ago which caused the formation of many conservationist bodies. My feelings are that now we need a controlled conservation policy set up in all areas of the nation to ensure that one species does not breed in sufficient numbers to cause harm to another. Of course, the biggest offender is man!

I am very aware of the many people who are against the killing of seals and deer, but I do think that these two animals are emotive ones. The seal cubs are lovely-looking, cuddly creatures and deer are regal animals with beautiful eyes. I don't think I could kill one myself and I do not agree with the method of destroying seal cubs by clubbing to death, but I do most certainly understand the need to cull these animals. It is all part of man's efforts to restore a natural balance. A conservationist once said to me that they are all God's creatures and have the right to live. I asked this question of him: if he found his home infested with rats and mice would he let *them* live in his house? They too are God's creatures, maybe not such nice animals but probably just as entitled to live as seals and deer.

In the mid-1950s on the side streams and main river on Broadlands Estate there were otters a-plenty. We would often see them playing with their cubs on the banks and on many occasions during the close season I would spend happy hours watching them. How much damage they did to our fishery, I don't honestly know, but I suspect very little. I certainly never begrudged them a fish or two. By the late 'sixties an otter was a very rare sight and we all missed seeing them on the water. Phillip Wayre, the well-known authority on otters, was speaking in Andover at a conference on the demise of the species. I discovered that he was in fact breeding them in the wild for release into areas that had previously had an otter population but now, for various reasons, held them no longer. After a discussion with Lord Mountbatten, he agreed with me that it would be a good thing to have a few otters on the Estate, and asked me to enquire of Mr Wayre about the possibility of buying a pair for Broadlands.

I quickly contacted the otter centre and they did have a pair for sale, but before parting with them they wanted to examine their proposed new home to see if it was a viable proposition. I met the representative at my fishing hut and we walked the whole length of our water. As we walked, he pointed out all the likely places for the otters to be released. There was only one problem, and that was disturbance by man. In the days when the otter was a common sight, our water would be fished on average by only three people a day. There was no road along the riverside, so very few cars ventured onto the banks. Most of the land bordering the river was down to grass for grazing cattle. Now there is very intensive farming, with bigger tractors and other machinery thundering about and many more fishermen patrolling the banks; cars, too, have free access on the tracks built alongside the river. Completely unintentionally we had made the water untenable for the delightful otter. The gloomy advice of the otter centre people was that it would be a waste of time and money to plant a pair on the river as, due to the disturbances, they would very soon leave the area. So to my eternal regret, there are no otters left at Broadlands.

There is just one exception, and only one. A company was involved in making a film of *Tarka the Otter*. Most of the filming took place around Stockbridge, some twelve miles from Romsey. For the production tame otters were used, and at some time, one of them decided to leave the set and disappeared. All the keepers were notified and asked to keep a look-out for the wayward animal. Only two days after his disappearance we had a report that an otter had been seen on our coarse fishing lake. He had swum from the island, clambered onto the bank between two fishermen, shaken himself and then wandered quite slowly over to the Blackwater, a river running off the Test, slipped in and been seen no more that day. He was fairly obviously the escaped star of the film as he showed no fear of humans. It was just an amazing thing that he had travelled so far in such a short time. Since then there have been isolated reports of sightings but unless he can find a mate, I'm afraid there will be no possibility of a return of these beautiful animals to our area.

In some places that have been left to nature, predatory species reign supreme. Some keeper colleagues and I had occasion to walk through one of the large enclosures in the New Forest when we were acting as marshals for the Oxfam Java Trail charity walk. It was a beautiful day and we stopped about in the middle of the enclosure for lunch. We all agreed that something was not quite right with the surroundings: it was the utter silence of the place. There was no rustling of squirrels, very little bird life, and it was just too quiet. Later in the day we talked to one of the Forest keepers, mentioning that we had found his enclosure eerily peaceful. He reckoned that what had happened was that the predators had moved into the area unchecked, virtually destroyed all other life in the woods and then moved out again. There is no doubt in my mind that a well-keepered area of land will hold more and healthier varieties of wildlife than many places left to nature and man in general.

As in the case of the otter on our water, the dramatic changes in the environment are causing most of the problems facing wildlife today. For instance, the river weed growths have undergone a great change in the last decade, as have water levels, flow-rates and colour, all having an effect on the life in the river. We now see weed that wasn't present a few years ago. Banks of starwort begin to hold up silt, then grow through it and so on until there is a solid wall of mud which is very difficult to shift. Blanket-weed now grows in larger amounts, smothering the life-bearing varieties. It is known that there is a movement of many thousands of tons of silt naturally down the river each year, and it depends on the flow whether this silt is carried out to sea or deposited elsewhere in the river system. It is beginning to look as if, due to the lower levels seen over the past few years, more and more silt is being deposited in the river rather than carried away. An example of this is that we recently dredged a short section to a depth of six feet in January and by April the water was only a few inches deep, the channel having been filled up again with silt.

The water nowadays stays coloured much later in the year, even without rain, the lower levels of water having a scouring effect and keeping the silt in suspension. The many new fish farms are combining with abstraction and forms of pollution from farmland and sewage to change the nature of the river. Even the fly hatches have changed: we now have stone flies hatching with the grannom, mayfly coming off for five or six weeks instead of the usual two, overlapping with the hawthorn fly, then long periods with no hatches at all. These changes must in turn have some effect on fish life that we may be just beginning to experience.

How do I feel about rearing fish and birds for release, to be caught or shot in the name of sport? I think it is more than justified if only on the grounds of the best form of conservation. As I have already said, good keepering preserves all wildlife. What would our rivers look like without the care, and indeed love, with which those of us fortunate enough to be river-keepers tend the waters and the life they hold? They would very soon be clogged with weed and silt, the fly and fish would suffer greatly and most probably the trout would rapidly be "fished out" and become a fish of the past.

What I do not like is the way trout fishing, and in some cases shooting is going. I call it the numbers and size game. I am not happy with the attitude of some of the new fishermen, where big is best and a lot is better. For many of today's fishermen a trout is not recognised as a fish unless it is over 2½ lbs. I am as guilty as anyone for helping to create this situation. I made the mistake of thinking, "If you can't beat 'em, join 'em." There is no doubt that a trout of 4 lbs is a very fine fish, but I feel it should be a real occasion to hook and land such a specimen. In 1957, during the time of mayfly hatching, Lord Brabourne went out fishing on our house beat. The fish were rising freely to the mays. One rise in particular looked like a big 'un just sucking the fly down. His Lordship cast at the rise, the fish took and charged off and was lost almost immediately. His Lordship went on upstream and successfully landed two fish of 2½ lbs and 2 lbs. Returning to where he had lost the big fish, he found it was on the feed again. Once more the large trout was hooked and lost. This performance was repeated once more when the fish came back to the feed. Finally a 4-lb breaking-strain cast was put on and again the fish was hooked and rushed off, fighting like a demon. At last it came to the net, a lovely wild brown trout weighing in at 8¾ lbs. It was an evening of a lifetime, with several fish over 2 lbs culminating in the "big boy", which Lord Brabourne had mounted and which is now on display in our main fishing hut.

How different today. The pictures of beaming captors, with an enormous fish hanging down over outstretched arms, I think look quite awful. The gentle art of fishing for smaller trout with seven- or eight-foot rods and 1½- to 2-lb breaking-strain casts seems in danger of deteriorating to the point of using salmon tackle for catching trout. Some waters already won't allow breaking-strains of less than 4 to 6 lbs. The only hope for the return to smaller fish and lighter tackle is price. The cost of trout food is ever increasing, which automatically puts up the price of each pound of trout reared, so there may well come a time when large fish will be so expensive to stock that owners and managers will have to increase their rents to a point where fishermen will not be able to afford to go fishing for the monster, carpet-bag rainbows. Then we will see a return to the gentle art and the strong, quick, good-fighting fish of 1 to 2 lbs, with the possibility of big-fish waters for those people who want to catch and pay for them.

With salmon, size is a different matter: the big, fresh-run fish is indeed a thing of great beauty, strength and fighting quality. But really it is largely a matter of good fortune, having cast your bait into the waters, whether a salmon of

40 lbs or 4 lbs takes it. I know that my personal best was a question of circumstances and good luck. It was early in 1966 and two of our regular fishermen had booked a day in the hope of getting a good springer. The date was 13 April – unlucky for some? They arrived as usual at ten o'clock. The weather was really dreadful, with high winds, sub-zero temperatures and sleet and snow on the wind. It was the sort of day which made the fire in our fishing hut very attractive. However, having taken the day, my two intrepid fishermen decided to have a go, then have a late lunch and pack it in. As they were only going to fish the beat for half a day, they suggested that I should fish with them to cover as much water as quickly as possible. I can't say I was exactly delighted with their invitation, for with the foul conditions outside it wasn't the sort of day I would have chosen to go fishing. But as I had no other fishermen out that day I was free to go with them. So, piling on as much clothing as we could, including waders, Barbour coat, balaclava helmet and gloves, we waddled off up the river.

Being the youngest of the trio, I was given the longest walk to the top end of the beat. I tried to think of a spot where I could lurk for an hour or so which would be a bit sheltered from the biting cold wind. I settled on the very top of our Lee Park beat by a may bush where I could crouch down and get a little respite from the weather. I had a 2½-inch Yellow Belly on my spinning rod, which I reckoned was as good as anything to fish with. I dropped it in under the bush and slowly retrieved, to find the line was freezing in the top ring. I had to keep dipping the end of the rod into the river to free the line and stop it freezing again. I was beginning to feel numb when on my third cast the bait stopped. "Bottom," I thought, and gave a tug.

To my amazement, I felt a double answering tug and with strong, powerful pulls, my line took off downstream. I galvanised my dead legs into action and lumbered over the bridge. By this time the fish had taken fifty yards of line and was still going downstream. I ran as fast as my bulky clothing would allow to get below the fish, which had continued his progress down river on the opposite side from me. He ran for about 150 yards, still keeping to the far bank, turned and ran all the way back to the lie where he had been hooked. At no time did he show himself, playing deep and strongly, absolutely out of my control, seemingly hardly aware of being hooked. I had by then been playing the fish, or rather he had been playing me, for more than thirty minutes, and still he was on the far side of the river and I had not had a sight of him. I was now considerably warmer, having trotted up and down for half an hour, and convinced I was into a foul-hooked fish, as I just couldn't do anything with him. I thought then that the best thing to do was to give him a bit of stick and take a chance of pulling out the hook-hold. The next time he stopped in his travels I began to pump him slowly over to my side of the river. The rod creaked and groaned and the line was as tight as a fiddle-string, but at last he was coming towards me. Finally he was under my feet but still very deep. My gaff was ready to use, tucked under my arm. I slipped it down the line, felt the lead, went further and then, feeling the fish, pulled the gaff home and lifted the fish. It was at that moment that I realised it was a big one. On the bank it looked huge, as indeed it was: 33½ lbs, and covered in sea-lice. We were all delighted, especially me, as I would certainly never have killed such a fish if it had not been for two rather mad fishermen wanting to go out on that impossible day.

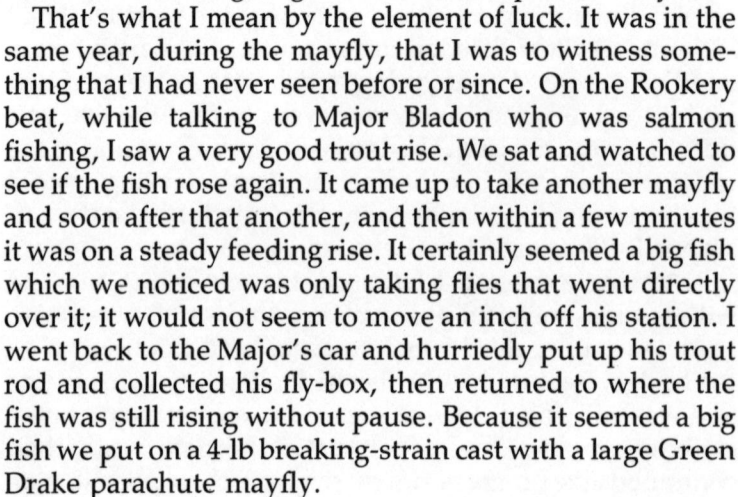

That's what I mean by the element of luck. It was in the same year, during the mayfly, that I was to witness something that I had never seen before or since. On the Rookery beat, while talking to Major Bladon who was salmon fishing, I saw a very good trout rise. We sat and watched to see if the fish rose again. It came up to take another mayfly and soon after that another, and then within a few minutes it was on a steady feeding rise. It certainly seemed a big fish which we noticed was only taking flies that went directly over it; it would not seem to move an inch off his station. I went back to the Major's car and hurriedly put up his trout rod and collected his fly-box, then returned to where the fish was still rising without pause. Because it seemed a big fish we put on a 4-lb breaking-strain cast with a large Green Drake parachute mayfly.

Major Bladon was a very skilled trout fisherman and cast the fly delicately, dropping it in exactly the right place. The beautifully presented fly floated right over the fish and was totally ignored. Our trout continued to rise, but scorned every expertly cast artificial put over it. I had a try and so did the Major's nephew; still no take. There was very little left in the fly-box we had not shown the fish, which was still feeding and yet was oblivious to our efforts, even refusing to be "put down" by some of my rather splashy tries. Eventually, in desperation the Major put on a "spent" mayfly, dropped it over the fish and was taken as soon as the fly alighted on the water.

Then the fun began. His eight-foot rod bent double and the reel screamed as the line shot out to the backing and continued to run with the backing line fast disappearing from the reel. All we had seen of the fish was a flash of silver. "I think it must be a sea trout," I said. But it didn't behave like one as it fought sub-surface, and the few sea trout I had played spent as much time in the air as in the water. After half an hour the fish showed signs of tiring and once rolled on the surface. It looked enormous. Luckily I had my large salmon net with me, for our tiny trout net would have been useless on such a fish.

At last the fish came round, the net was under it and it was lifted safely onto the bank. The fish weighed 9 lbs and looked very much like a salmon to me. When I opened the fish we were able to identify and count 120 mayfly in its stomach with a lot more we couldn't separate or even identify. There was a big discussion and quite a controversy in the fishing hut as to whether the fish was salmon or sea trout. The river board biologist read some of the scales I had taken from the fish and gave a positive identification: it was a salmon without doubt, and with photographs of fish and stomach contents a report was sent and published in the *Fishing Gazette*, causing a great deal of interest.

In the 'fifties and 'sixties we did kill a few sea trout, mostly on salmon flies. No-one really went fishing for them at night, so a true evaluation of the sea trout population wasn't possible. On average two dozen of these lovely fish were killed each year on Broadlands water, all very splendid fish. The best one I saw was 16 lbs. I do wish a good run of sea trout could be established on our water. Although a great many do come into the river, and in fact get caught in the lower reaches, once they start their journey upstream they come so far then turn left up a small river, the Blackwater, which is a very acid stream taking much of its water from the Forest land. It would seem that sea trout like the acid rather than the alkaline water of our chalk stream. What I don't understand is why the few fish that do run up to the higher reaches of the Test have not spawned successfully and over the years developed into a good run. Is it just that fish that are born in the alkaline water will not return to that water? Or is there some built-in preference for acidity regardless of their birth water? And why do some fish run into the alkaline water in the first place? There is so much we don't know and in many ways I hope we never discover, for it is the mysteries of fish and fishing that are so fascinating and keep us piscators going.

CHAPTER TWELVE

THE TEST DOES not only contain fish. In the past we have made many interesting discoveries in the river, and many things have fallen into the water. One of the fishermen had shown me very proudly his new wrist-watch, a solid gold Rolex with a gold band. With his second cast of the day the watch flew off his wrist to land midstream in about twelve feet of water, never to be seen again. A stolen one-armed bandit was dumped in at Middlebridge and could be clearly seen, surrounded by sixpenny pieces, lying on the bottom. I found a splendid brass, very old Austin hub-cap nut, which I have since cleaned and polished. But a really strange occurrence was when a chap came puffing up to the fishing hut to tell me we had a large bomb in the river. I put on polaroid glasses and went to have a look. Sure enough, lying in about six feet of water was an object that looked like a bomb. I returned to the hut and phoned the police. They too had a look and confirmed it as a bomb of some sort.

The Army bomb disposal squad was summoned and a team of six men arrived. They put on their diving suits and we all went to examine our bomb. The officer in charge carefully slipped into the river and very gently measured the length and diameter of it, consulted a book and came up with the answer. It was, he said, the exact dimensions of a sixteen-inch naval shell, although how on earth it had arrived on our number 3 beat was a mystery. We were all sent away to a safe distance while the Army chaps rigged a crane to lift the shell from the river. As they were scooping out the river bed under the shell to pass a chain round it, the shell suddenly moved and rolled a little further into midstream. A sixteen-inch naval shell would weigh something over a ton and would not have moved easily, so two of the men took hold of it and to their surprise lifted it up and out of the water, onto the bank. There they found the shell to be made of wood – it turned out to be a sawn-off pointed end of a river pile. The officer asked if they could take it away for their museum as it was in fact the precise measurements of the shell they thought it was.

The most intriguing discovery I have made was during a spell of very low water level and a crystal-clear river. A post projecting from the bank was exposed. We knew it was there and it had been a constant concern of mine in case a hooked fish should become snagged on it and lost, so I took the opportunity to put on waders and, using the chain-saw, cut it off close to the bank. Having removed the offending pile, I realised that the river bed I was standing on felt springy and not at all like the gravel bottom it should have been. I dug about with my foot and could feel a ledge. I lifted it up and found it was a plank of wood, sixteen feet long by eighteen inches wide and two inches thick. When we managed to get it onto the bank we found some square-headed nails embedded in it. They were obviously very old and hand-made. I investigated further by raking away some of the gravel bed, discovering more timber planks, but these were still firmly nailed to beams running at right angles to the bank. Further scratching about revealed an enormous baulk of wood eighteen inches square and thirty feet long. I swept it off with a broom and found, let into the baulk, equally spaced large tenon joints, each one about a foot long and three inches wide.

Bearing in mind that the large piece of wood was only discovered because five feet of bank had collapsed, I realised it must have been lying there for donkey's years. I telephoned our local archaeological society, who were very interested and sent out two of their leaders. They thought it could be a Viking boat. This caused quite a bit of excitement and it was decided that a lot of careful excavation should be carried out. At the time Mrs Margaret Rule was in the process of recovering the *Mary Rose*, the Tudor warship, from the Solent. She agreed to come to Broadlands with her diving team and make a survey of our boat. They came, and over two days cleared away the silt and gravel, exposing as much of the structure as possible. It was carefully measured and eventually a drawing was made. We named it the "Thing", for in some aspects it resembled a boat, with ribs and planking which could be a hatch-cover or decking. Another opinion was that it was an old building which had collapsed into the river, or possibly an old bridge. The bridge idea was discounted due to the very steep bank on the other side of the river.

Lord Mountbatten became very interested in our find and asked an expert on ships from the National Maritime Museum to visit the site and give his opinion. In his view it bore no resemblance to any boat he had ever seen, but he thought it might be a specially built stone barge, used to transport stone to build Romsey Abbey.

After the first flurry of great interest, all enthusiasm for the "Thing" evaporated, and no further examinations were carried out. What amazes me is that no attempt to carbon-date the beautiful hand-made nails was ever made, because whatever else our "Thing" is, it is very old and a mystery which is still lying unsolved on the bed of my patch of the Test.

There have also been some very strange things hooked by our fishermen. Walter Geary was fishing just above the hut when he got stuck into what he thought was quite a big fish. The fish behaved in a peculiar manner, running strongly downstream with a strange diving motion like a switchback. Walt had very little control over it but slowly eased it to the side by walking downstream and putting on a lot of side-strain. Finally I thought I could see a dorsal fin, and swiftly slid the gaff over and pulled. To our amazement, the gaff met something hard and skidded off. I had another stab with the gaff and again felt a hard resistance but this time hooked it and lifted out a complete skull of a deer with six-point antlers! We took our trophy and fixed it above the porch at the fishing hut, where it hung for many years until one night the place was broken into and vandalised and our antlers were smashed into a thousand pieces.

This sort of thing is one of the new problems we have to contend with on the river. I just do not understand the mentality of people who have to destroy. We have had many instances of vandalism in the last few years and it appears to worsen as time goes on. Our stews were broken into and the asbestos roofs of the food and equipment stores smashed in. Having used our nets and waders to take a few trout, they then slashed the nets and waders, broke the handles in two and proceeded to throw everything movable, including our wheelbarrow, into the stews, killing many more fish. Fences were pulled down, screens removed and the whole place looked as if a herd of elephants had been rampaging about. All our hard work was demolished in one night. I feel very bitter and vicious towards the mindless thugs who carried out the wrecking process. On one occasion I caught six young men jumping up and down on one of our narrow bridges over a stream. They were trying hard to break it. We have had noticeboards uprooted and thrown into the river. If only we could harness the energy these chaps use to destroy, and use it for building, I don't think we would have any troubles with labour ever again.

This is one of the more unacceptable changes that have occurred during my time at Broadlands. One does, of course, expect things to alter in a period of nearly thirty years. The salmon runs have changed dramatically but the timing of the fishing season has not. Where we used to find fish in January, February, March and April, it is no longer so. Although our season officially opens on 17 January, the spring run has declined to such an extent that we at Broadlands do not consider it worthwhile to begin fishing until 17 February, and really, if the past three or four years are an example, 1 April would be soon enough to start. However, whereas June used to be our best month, now we don't start to have many fish until the last few months of the season, July, August and September. Indeed, since the later 1970s the river holds a great many fish after the close of fishing. These are not gravid old red kippers, but the majority are fish one would be quite happy to catch with a clear conscience. I feel that it would be a good idea to extend the fishing, in common with trout and sea trout, until the end of October, and start the season a couple of months later. This could give the few spring fish a chance to run up unmolested and may help to re-establish the big fish on the Test. The problem is that to alter the season does require a lot of paraphernalia, ending by having to go to Parliament. A much better idea would be to give control of opening and closing dates to the area river authorities, enabling them to make the season flexible according to the salmon runs on the individual rivers. I hasten to add that I am not advocating a willy-nilly taking of fish ready to spawn. I would really like to see a two- or three-year trial under strict conditions and close observation. We could, for instance, make the last month of the season fly only and have a "no gaff" rule, and all hen fish returned to the river. All I want is to try and improve the catch of fish for our fishermen, but not to the detriment of the salmon, for they are a great part of my livelihood.

In the last few years the methods of fishing have seen great changes. Until recently my own rods were all of built cane, and I resisted any other material. I had tried fibre glass fly rods but I thought they lacked the feel of cane. I took my thirteen-foot Sharpes Splendid rod to Scotland and, after two days of twelve hours' hard fishing each day, discovered muscles I didn't know I owned! A friendly fisherman loaned me his very splendid fifteen-foot carbon-fibre rod. What a difference! Lighter by far and very powerful. I became a complete convert and, although I still use my cane rods on my own smaller river, for long casting on big rivers with the large weighted tube-flies, carbon is the one for me. I have used rods made from Boron, which I found lacked power and seemed a trifle soft and slow in action.

The other thing with these new rod-making materials is that they can almost invariably take a heavier line than recommended by the manufacturers, and indeed work better. Regarding fly lines, I still, deep down, prefer the old dressed silk Kingfisher-type double taper. These lines, although a bit of a bother to grease and dry after use, are very good to cast, especially in high winds. I now use a sink-tip plastic line, also double taper, much preferring them to the weight-forward lines. A point of economy arises with the double taper. They can be reversed after a few seasons, extending the life of the line considerably.

Flies, too, have undergone a change, hair largely taking the place of feathers, and they are also much more lightly dressed. I remember when we bought flies in the old days, Walter always said they needed "plucking" before use as they were so heavily dressed. It's a great pity that there is a restriction on the import of some feathers, especially Jungle Cock. I do like to fish a fly with Jungle Cock cheeks – they seem to show so well in the water. The substitute doesn't look half as nice, but that is just a personal opinion, and anyway, having confidence in the fly or bait you are using is halfway to catching fish.

There is so much rubbish written about fly-dressing. I don't think it really matters whether a salmon fly should have a feather taken from the wing of some obscure Chinese bird or not. I remember finding a fly caught in a tree, lost the previous season by an unfortunate fishermen. The fly looked suitably scruffy, so I thought it worth a try. It killed three fish that day and created a lot of interest and argument as to which pattern it was, because all the other fishermen wanted to buy some of such an obvious killer. I sent the rather tatty-looking fly to Hardy's, with a request to identify it and possibly tie some for me. Their reply was, "It's a Silver Grey, with all the essential feathers missing!"

One of my most successful flies was one I tied using a

vivid blue feather from my daughter's toy shuttlecock, and a blue wool body taken from a sweater my wife was knitting. It bore no resemblance to any "proper" fly-dressing but it killed a lot of fish before it finally disintegrated.

Something similar occurred in Scotland. I had spent many hours during the year tying "Garry Dogs" on two-inch brass tubes, carefully making the black body with silver tinsel winding. Marie and I set off for Inverness with what I fondly imagined to be plenty of flies, but I took my fly-tying kit just in case. After a few days' fishing in howling winds and rain, my fly stock was diminishing at an alarming rate. It reached the stage when I was simply tying flies at night for the following day. Because of lack of time and some of the proper material, I finished up by just lashing the dyed hair onto the undressed tubes with dental floss! The resulting "flies" didn't look very beautiful or professional, but they too killed a lot of fish.

On my own water we used to kill a third of our total catch on fly, but now very few are taken by this method. It is, of course, directly a result of the ratio of fly fished. Fewer fishermen now use fly, probably because it is easier to spin using the modern reels. It's a great shame that fly isn't used more as it is a most efficient way to cover the water. However, spinning is fun and certainly gets fish, as do prawn and worm.

The prawn is a much-maligned fishing method. I have heard and read of dozens of fish leaving a pool, never to be seen again, once a prawn is fished through it. I must say I have never witnessed such an event. I have seen fish move away from a prawn as if scared. I have also taken the same fish later in the day. In fact I had a tussle with a fish which I would find hard to believe if it hadn't happened to me, and was witnessed by one of my fishermen. It was at the top of Rookery Pool, where I had stopped to chat with David Le Coyte who was fishing on that beat. We were standing close to the river and as we talked I was, through habit, peering into the pool. There, only a couple of yards away, was a fish of about 12 lbs. We were afraid to move in case we scared the fish away. David gave me his rod with a prawn mounted, and I gently lowered it to the fish, which moved quietly up and took it. I struck and the fish came to the surface, threshed about, came off and swam in a circle, back to where he was originally lying. On examination, the hook had straightened out.

As the fish was still there we put on a new hook and tried again. Once more, as soon as the fish saw the prawn he took it. I struck, two thrashes on the surface and gone! Another straightened hook. The fish repeated his last performance, returning to the same spot.

Another new hook and prawn were offered and taken at once. Exactly the same thing happened. The fish came onto the top of the water, wildly splashing for a few seconds, and was lost. I couldn't stand it: yet another hook straightened. They seemed as if they were made from paper clips! The fish, meanwhile, had returned to his lie.

I walked back to my Land Rover to fetch one of my own prawn tackles, having said rude words about David's. When we got back to the pool we found the fish still waiting for us, so once again I gave him a prawn. It hung in front of him for about half a minute, when he began to drop back, then suddenly he came forward and took the prawn really savagely. This time there was no mistake, the hook held and after a fierce struggle, David netted the fish, which weighed 11 lbs. That fish had been hooked and lost three times. On each occasion it was well on with a tight line and bent rod, yet it still took a fourth time, and at no time seemed to be worried by the prawn. There is no doubt that it was the same fish as it was in our view from start to finish. It was certainly a rare and bewildering experience.

When discussing prawn fishing, I am always asked the hoary old question, "When do you strike?" I don't think that any general rule can be applied. I have heard people say that you must wait for the "third knock", which is ridiculous. The only thing I can say is that I strike when I feel that the fish has really taken it. Sometimes I have seen a fish take a prawn like a ton of bricks, struck and not even felt the fish. Conversely, I've had them gently plucking at the bait for minutes on end, finally striking and finding the fish so well hooked that it would never come unstuck. So prawn fishing isn't the easy way to catch fish that many fishermen who don't use it think it is.

The tackles used for mounting the prawn doesn't seem to matter very much, as I am sure a fish taking a prawn doesn't stop to consider whether the prawn is spinning, wobbling or just moving up and down. It is largely the presentation of the bait to the fish and the depth at which the fish is lying. So many fishermen do not use enough lead or alter their weight according to the water being fished. I have used the old Berrie Mount, which is no longer available, the spinning mounts and the sink and draw type. In earlier years the up-trace spinner was a popular mode of mounting prawns. It consisted of a wire trace with the propellor at the top and well away from encumbering the prawn which was attached to the bottom of the trace. Wire had to be used to transmit the spinning action from propellor to prawn. I was nervous using this method, if a loop or kink should be thrown into the wire trace in the course of casting, it would break like a piece of cotton once any strain was put on. I now use a pin with an eye twisted onto the front, made from a straightened paper clip, pass the nylon trace through the eye, tie on a treble hook and secure it to the prawn with copper wire.

It is more important to be particular over the prawns themselves. I prefer to use fresh English prawns which are smaller, firmer and a better colour than the usual ones bought from the fishmonger. These, more often than not, are Norwegian, soft and pale in colour, too big and very much inclined to break up after fishing with them for a short while. I'm not very keen on using prawns that have been dyed, especially the rather violent purple colour. I realise that these will take fish, and on some rivers in Scotland the ghillies swear by them, which is fine. I would not presume that anything I say can apply in a general manner to all fisheries; this is just my own experience on my water. Different rivers, different methods.

This is why local knowledge is invaluable to fishermen going to new waters. It pays to try the proven methods of fishing even if it goes against the grain. If they don't work, then have a go with your own ideas, for if you aren't catching fish anyway, you may as well not catch them using your favourite methods. And who knows? You could come across a fish that doesn't know the local rules. Dear Old Walter taught me his maxim for fishing, which is "Never say never, and never say always". For you can bet your life that if I say you will never have a fish on a particular fly or bait, that will be the one which takes a fish.

Worm fishing has been re-introduced at Broadlands in the last couple of years having been banned before. We do kill a large percentage of our annual catch on them, but again, it is relative to the ratio fished. There is little doubt that fishermen come into my fishing hut, look at the records, see a fish killed on the worm the previous day and think that it is their best chance of a fish. So they use it, catch a fish, and so on. There is no question that to use a worm requires a great deal of skill, that is, to fish it consistently well. I find, as a newcomer to this mode of fishing, that it requires a lot of self-control. With the rod transmitting knocks and bangs while I am allowing the fish to take it properly and indeed giving line, to prevent myself striking is a very real problem. The worm also presents another question relating to the feeding of salmon in fresh water, for some of the fish I have seen taken on worm have without doubt taken the bait well down into the gullet, beyond the point of no return, which does point to the fish eating. I have even seen a fish which has taken the worms and passed the crushed mass out through the gills; when it was finally netted, the hook was outside the gill-covers. Do they eat? Never say never!

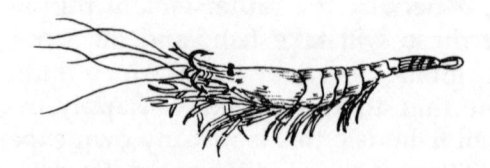

CHAPTER THIRTEEN

WHEN LOCAL RADIO STATIONS were started in the early 1970s our own Radio Solent decided to feature a fishing programme which they called "Tight Lines". John Piper was the presenter and I was asked to go along to the studio to record a programme on the Broadlands fishery. John and I became firm friends and with Robin Worman, one of the other BBC presenters, made several more recordings on fishing. John is a well-known fisherman and writer and is very keen on sea fishing. Through him, I too found it interesting. After many chats on the subject we formed our own Radio Solent sea angling team of six. We hired a boat with Eddie Wood as skipper and had many happy days in the Solent, catching fine skate, cod, conger and my particular favourite, bass. We changed our venue to go out to the Needles, which is quite a long trip but was usually worth it as the fishing was so good.

What I find most fascinating with sea fishing is that you never really know what sort or size of fish will take your bait. I do know there are some massive fish lurking in the deep water off the Needles. We were out there a few years ago and had been having bites galore but bringing only a few fish into the boat, when one of the chaps had a really good solid take. The practice was to let the bite develop so that the fish had plenty of time to get well hooked. Once he did strike, the rod bent well into the fish, but no movement was apparent. The skipper was of the opinion that he was hooked into a deep-sea cable, as every time a few feet of line were recovered, the sheer weight on the end of the line took it back to the sea bed. Then very slowly the line moved out and described a circle of about forty feet. It was certainly a fish, but what?

By then half an hour had passed and the fisherman's arms were aching, so he handed the rod over to another member of our party. There was nothing we could do with the fish. He lay dormant in spite of the steady strain put on him by a powerful boat-rod and 50-lb breaking-strain line. We held a council of war and decided to try a kill or cure method to get the fish off the bottom. One of us was to crouch down with the rod resting on his shoulder, then gradually stand up, thus recovering line. This worked well for a time and twenty-five feet were gained. But the fish then used its power and almost effortlessly swam back to the bottom, the clutch on the reel getting quite warm! This performance went on for an hour and three-quarters before the hook-hold failed. We reeled in, to find a piece of flesh on a very bent hook. We had not even had a sight of the fish but it must have been huge, probably a mighty skate.

While bass fishing one day, Norman Falconer, a very experienced fisherman, got stuck into a large bass. We all reeled in to watch him play it. Then I saw the bass and to our amazement, it was in fact a salmon of about 14 lbs. There is no doubt it was a salmon; we all saw it at close range before it, too, was lost. I often wonder what prompted that lone fish to take a bait at sea, and if any other salmon have been taken by sea anglers on rod and line, and not reported.

Now our Solent sea team has broken up and I do very little fishing at sea. I still prefer to eat salt-water fish. I'm certainly not very fond of trout, although I have said I am on one occasion. I think it was a justified white lie. I had been approached by the Cedar School, Southampton, to talk to the pupils about the river and its life as they were doing a project on the Test. The Cedar School is a school for disabled children. Having spoken to these lovely kids, I thought it might be possible to bring them to the river actually to see all the things I had described. The school has its own mini-bus specially equipped to carry wheelchairs and, when I showed its driver the sort of terrain he would have to travel over, he very gallantly said he would have a go. It was decided to bring the children in two groups on separate days. The great day arrived and after a rather bumpy journey we began unloading the wheelchairs, lining the children around the stew ponds. I gave them some fish food to throw in for the fish. The excitement was intense as the fish splashed about feeding. The children loved it and didn't want to go back to school.

It was the same the following day when the second group came, except that before they boarded their bus they all gathered round me to say "thank you". Then two of them trundled forward in their wheelchairs and asked me, "What is your favourite food?" I had already been warned of the question by the bus driver, so had my answer ready. "Rainbow trout is what I like best, especially stuffed rainbow trout." Their little faces lit up and they then produced a carrier bag each which they handed to me. In the first bag was a beautifully painted cut-out fish and in the other, frondy pieces of green and brown paper carefully threaded onto a length of string. This, I was told, represented the weed in the river. The fish and weed had been painstakingly made by the local disabled children, and the fish is hanging in our main fishing hut as a poignant reminder of those two very happy days. More recently, schools in the area have paid us visits. From one school I had nearly 100 letters of thanks, every one a gem, which I treasure.

Of all the things the children saw on their visits, the best was a ride on a rather smelly tractor and trailer which was used for silage carting! I thoroughly enjoy working with young children and am constantly surprised at some of the facts that stick in their minds.

I've had fun making radio broadcasts, but was rather nervous about television. My first experience of filming for TV was with Martin Muncaster, who guided me firmly but naturally into the film world. He always insisted on everything being done as it would be in a normal working day, and I think this comes over in his programmes. In 1978 I was approached by Michael Croucher of BBC Bristol to make an hour-long film to be entitled "A Year in the Life of a River Keeper", one of a series being made.

The plan was that a film crew would visit me throughout the year, filming whatever was going on at the time. Lord Louis's permission was asked and readily given. He even agreed to take part, and he didn't miss the opportunity to put in a little light advertising for the opening of Broadlands House to the public! The film crew were a very happy bunch who quickly put me at ease, for it was quite difficult to carry on as normal with the lens of a camera following me around, and not to look at the camera. We had formed a programme for the year, with all the important happenings to be highlighted, like weed-cutting, mayfly time, stocking the trout stream, catching salmon and trout.

We began with the opening day of the salmon season. All went well, and periodically I would phone Michael to let him know when we were doing anything of interest. The mayfly were beginning to hatch and I thought it best to wait for a few days before arranging the filming, to let the mayfly hatch really establish itself. Having had a good show of mays for a week in fine weather, the TV crew arrived to a cold, windy, wet day and, of course, no mayfly. Instead of wasting the day, Michael decided to film some indoor chats, so I was "taken" sitting at the hut table making up the game book and later in the day we had a very hilarious tea party. After tea the sun broke through and we rushed upstream to see if the mayfly were coming off. Talk about luck! It was one of the best hatches I have ever seen. They were coming off in clouds, reels of film were used and one shot showed me standing watching a mass of dancing mayfly over the trees and my back covered in flies. The trout were going mad, feeding ravenously on the hundreds of flies on the water. Michael wanted film of a trout being caught, but I didn't think there was much chance of me doing it. However, I put up my rod, tied on a large Green Drake and cast it out amongst the great crowd of naturals already on the surface. The fly alighted and was taken at once. It all happened so quickly that there had been no time for the cameraman to film the take. He did get me playing and landing the fish. I hadn't worried about being filmed catching a trout as I thought that we could do a mock-up if need be, but Michael would have no part of it. The whole film was to be actual with no faking at all, and that is exactly how it was: no rehearsals, first-time takes, as it happened at the time. The one thing that I couldn't and didn't do was to land a salmon to order. Goodness knows I tried, but it seemed that whenever the TV crowd were with me, the fish knew and would not co-operate, so the film lacks a salmon kill.

As a part of my year they filmed our daughter Mary's return home from Australia. This was real fun as it was a complete surprise for Mary. We had to leave for the airport very early in the morning so we sent a front-door key to Michael to enable him to go into our house and set up his lights and cameras. Poor Mary didn't know what was going on when she arrived home to find the inside of the house brilliantly lit as if with tropical sunshine. Later that year the cameras were present at her wedding and the reception afterwards. Lord Mountbatten had kindly allowed us to hold the reception at the Orangery, a very lovely setting on a gorgeous sunny day. The film is a super record of the ceremony and a copy was given to Mary as a wedding gift from the crew.

The film was completed within the year and was shown nationally in the following May. It was a great success and I had quite a lot of fan mail, asking for more programmes of a similar nature. Among the letters I received was one from a long-lost aunt who had seen the film and another from a cousin I didn't know I had. I also received a parcel containing a book. The book was a copy of *Salmon Rivers of*

Scotland by Augustus Grimble, very kindly sent to me by a lady whose father was a great fisher, and who had noticed that I had read a quotation from the sister volume by Grimble, *Salmon Rivers of England and Wales*.

It was a strange experience, seeing myself on TV with all my friends and workmates. I've watched the film several times since and don't think I will ever get used to it. On the question of fishing films, I have spoken to TV producers on the subject and we all agree that to make a good film that is to please fishermen is a very difficult job. If it shows fish being caught all the time it rings totally untrue, but who wants to watch people fishing all day for a blank? Although blank days are much more usual than fishy ones, they don't make for good viewing. So although fishing is one of the biggest participant sports in the UK it is not very well represented on the media.

In 1981, a fisherman friend, Ken Woolfitt, asked me to keep a date in May clear as we were to go out for dinner in the evening. This was to be a surprise "do" for another fisher friend, Catherine Henderson. On the night, Ken and Catherine arrived to pick us up and away we went. I didn't know where we were going to dine and was a bit mystified when we turned off towards the fishing hut. Ken told me we were taking a short cut through the Estate, which I accepted as a reasonable explanation. As we were about to pass the hut Ken said, "We can call in to see if there have been any fish killed today." He stopped the car in the compound and I got out and went into the hut. I was amazed to find it full, but full, with fishermen, with the table laid with food and drink. I was absolutely staggered to learn that this was to be a party to celebrate my twenty-five years at Broadlands.

For once in my life I was completely speechless, as I had no idea of what was on. The whole thing had been secretly organised by Ken and Catherine, involving a lot of hard work but, they have told me since, a lot of fun. It was a great evening, a memorable and, indeed, moving one for me, seeing all my old friends and culminating in the presentation to me of a cheque and an engraved statuette of a fisherman holding a salmon. I was overwhelmed by the generosity and good wishes from my fishermen and colleagues, and went home still a bit dazed by it all.

I must say that those twenty-five years have passed so quickly it doesn't seem possible. I only hope that the next twenty-five do not go as fast, but I fear they will.

Looking back to my early days at Broadlands, we were a team of four: Walter, Len, Sid and me. All we had to look after were four salmon beats, and we did very little paperwork. Today there are six salmon beats, three miles of trout stream, trout rearing and our twenty-six acre coarse fishing lake, and we are still only four men: Brian, Mark, John and me. Brian Parker is my second-in-command and came to Broadlands in an almost accidental way. I had been asked to give a talk to the Romsey Young Farmers Club. At the time, I had a vacancy on my staff due to one of the old hands leaving. When I had finished my talk, there were lots of questions and there was a deal of interest in my work. So I said, quite jokingly, that if there was a keen young chap in the audience looking for a change, I would give him a job. I was approached afterwards by this young lad, who said he was a baker but was anxious to get out into the open air. The net result was that I took him on, and now Brian has been with me for over ten years, is a dedicated keeper and fisherman and, I am sure, will one day take on "head boy's" position and carry it off in the best possible manner. He is, incidentally, still a very good confectioner, for he iced both my daughters' wedding cakes with a perfection that would do credit to any professional of many years' standing.

Mark Simmonds I met as a student from the Hampshire College of Agriculture while he was on a course studying river-keeping and fishery management. He was primarily interested in coarse fishing, so when we opened our lake he was the natural choice for the job of running it. He came to Broadlands just as the lake building had been completed, and has worked on it tirelessly ever since. It is due to his devotion to the job and the many hours over and above normal working time that he puts in that the lake is a success and improving every year. But this is really what keepering is all about. It most certainly is not a nine-to-five job or a question of clock-watching. We are lucky to be doing work we love and derive a lot of job satisfaction. In fact we mostly look after our waters as if they belong to us – although we don't object to our employers actually owning them, they are really ours!

John Dennis is the youngest member of my team, who came to me as a student on the Work Experience Scheme and stayed on when a vacancy occurred. He too is a dedicated lad, works with a will, and is without a doubt a very good keeper in the making.

I am very fortunate in having a good team working with me, who in spite of the extra work-load still manage to keep our patch in order and looking nice. Of course, we now use some machinery for bank-trimming and the like, but jobs like weed-cutting have to be selective and are best done by hand. One piece of machinery we use to help us in our efforts to maintain a natural balance of fish species is our electric fishing kit, which to begin with we used ad lib to remove unwanted fish from our waters. However, it was noticed that after we had electric-fished a stretch, a considerable number of fish of one or two inches long were floating on the surface, dead. On examining these bodies, the cause of death was not apparent, and it wasn't until the day they were X-rayed that it was discovered that the majority had broken backs. This had happened when the small fish, on receiving the electric shock, had gone into a violent spasm, thus fracturing their vertebrae. This raised the question of whether the predators removed from the river did more damage than the electric fishing.

The deaths of these fry also set me wondering what harm, if any, could be done to snail, shrimp and fly life, together with the millions of microscopic organisms so very necessary to maintain the existence of other aquatic life. There is no evidence of such damage, but I do feel that some harm could result, so only use electric fishing very sparingly. It is most effective on our trout stream, which is narrow and not too deep, but doesn't work so well in the wider, deeper main river and is only rarely used there, to remove a particularly difficult pike from a known site.

Like sea fishing, the use of electric fishing is very interesting as you never know what is going to turn up. We have even discovered flounders, a sea fish, some three miles upstream in fresh water. It seems that the bigger the fish, the quicker they succumb to the "electric net", most likely because of the greater length of lateral line, the fish's nerve centre, exposed to the current. It doesn't appear to harm larger fish, although I think the shortest possible time should elapse when the fish are exposed to the current.

We never kill any fish we want removed from the stream. They are all taken in tanks to our lake to supplement the stocks, even the grayling which, despite the general opinion that they require flowing water, do very well in the comparatively still water of the lake.

I would stress that the use of electric fishing equipment is a job requiring great care and should never be undertaken by inexperienced people. The kit, too, must only be held under licence from river authorities. It can be a very dangerous game, and I consider myself lucky in getting away with only nasty shocks in the early days of using this method. On one occasion, I was wading a side stream holding the electric net and having removed a pike, threw the net onto the bank and began to clamber from the river. As I did so, I slipped and my hand came into contact with the electrified mesh of the net. I went absolutely rigid and thought I was a goner. Fortunately, one of the lads was near enough to the generator to cut the power off, so all I suffered was a severe shock and a hard lesson learned. When electric fishing now, we always have a man standing by the generator ready to shut off the current if any emergency arises.

Never allow any member of the team to put a hand in the water when the electric net is immersed, for if he does he will be in for a surprise – or rather shock! The modern, pulsed electro-fishing equipment is certainly much safer to use. The net has a "hold-down switch" on the handle which automatically cuts off the power when the hand holding it releases pressure. I must say that I am usually very surprised at the number of fish that are in the stream. I would like some of the fishermen who tell us "there are no fish in the stream" to be present when we electric-fish.

For my own part, my job has taken on a great change. With the growth of the fishery I am now more involved in the actual running of it, which includes the financial aspect. While balance sheets are still a bit of a mystery to me, I think it gives a greater insight into the economies of fishing, and I am constantly aware of the ever-rising cost of maintaining the river. Fishermen may think that the rents they pay to fish are on the high side and often say it would be nice to have a new groyne here or a platform there and it would be convenient to have a bridge over the river in a certain place. I agree with them until the costs are worked out! The timber alone for one breakwater, even a comparatively small one, is priced at around £200, and as for steel piling, I could easily use up three years' maintenance budget on protecting thirty yards of bank.

The banks of the river these days take a bigger hammering than they used to, due, I am sure, to the more rapid rise and fall rate after heavy or prolonged rain, and the silting that is taking place. There are two pools on the river that have shown an incredible change. Cowshed Pool, when I first came, was situated in a large curve, where the water swept round in a great whirlpool. Slowly over the years the river eroded the left bank and deposited silt on the right. Eventually, we had to build a long catwalk some five yards out to enable the pool to be fished. In a few more years, due to continued erosion and depositing, a new catwalk had to be constructed five yards further out, then yet another was made an even greater distance out into the river. Eventually we were obliged to pile all along the right bank, a distance of two hundred yards. The cost was astronomical, even using wooden piles grown on the Estate.

It was amusing when the Ordnance Survey people sent out two lads on an exercise to plot the course of the river, a job that hadn't been done for many years. They did their survey, but a few days later returned accompanied by two older men, who went through the whole procedure again. The very next day a mini-bus load of senior surveyors came to the river with the lads, convinced that they had got it wrong, but were startled to find that at one point on the river, it had in fact "moved over" a distance of nearly forty yards in a period of only twenty-five years.

Another pool which was dredged to a depth of six feet in an attempt to increase the flow of water and prevent silting, had just the reverse effect, resulting in a groyne that used to produce quite a few fish is now high and dry, five yards back from the river on solid ground. These happenings are just two of the things we try to contend with as part of the maintenance programme.

With these changes in contours there are inevitably alterations in the old salmon lies, so we are constantly looking for ways to improve the water's holding capacity. Some years ago I read a paper written by an acknowledged expert, suggesting that to create places attractive to salmon, blocks of concrete made in various shapes should be dropped into the river. We followed these plans to the letter, building blocks pear-shaped, triangular and in the form of a pyramid. We made a total of twenty of these which were carefully lowered into places where we had not previously had fish residing, as I do not believe in messing about trying to "improve" places which already hold fish since there is a very real danger of ruining a pool in this manner. The concrete block idea seemed a good one, but I must admit that it had only a very limited success, for we only ever found fish behind one out of the twenty we put in. However, we decided to try it again, so we made more blocks, this time three-foot cubes, and again put them into non-productive places. We had a terrible struggle getting them into the punt as they weighed a great deal, but we managed and they were all duly launched – they are all still sitting on the river bed, waiting for salmon to take up residence.

I have to be very aware that whatever work is done in either river or bank improvements, the direction of flow can be slightly altered, with the possibility of affecting a pool quite a long way downstream. It is for this reason that a great deal of thought and discussion goes on before installing a new groyne. I learned my lesson the hard way, in Walter's day. He thought to enhance our Iron Piles Pool by putting in two small breakwaters equidistant along its long length. They were very tiny groynes, but the water flowing round them must have had a tremendous scouring effect which caused the undermining of the steel piles and the bank to subside behind them. The piles were of the interlocking type which meant that they could not fall into the river, as unsupported timber ones would have. What happened was probably worse. They stayed together but bowed out in a large bulge, completely altering the flow direction, and they had to stay like that for months until the specialised equipment required could be obtained to remove and replace them with new, longer piles. By then the damage had been done. The pool used to produce a lot of fish; I had seven in a morning on prawn, before the disaster. These days very few fish are taken there, even though it has been restored to its previous contour and is a very good place to hang the prawn as the fish lie tight under the bank.

It was along these piles many years ago that a fish was taken on a bait that Walt and I had never seen before and which Walt said no self-respecting salmon would ever look at. It was a bright pink Kynoch Killer, and of course a 25-pounder took it at once. Another prime example of "Never say never"! Over the years I've seen fish killed on all sorts of weird and wonderful baits. We had a marvellous character living in Romsey known as Professor Woodley. He was a magician in every sense of the word. A member of the Magic Circle, he would entertain us at tea by making spoons disappear and producing shillings from ears of fishermen, causing a lot of enjoyment. He was also a magician at catching fish. He reckoned that salmon would take anything if it was presented properly and, most importantly, the fish wanted it. His own favourite for early fishing was to buy a pair of kippers and cut a piece of the skin the shape of a fish, then mount it on a spinning tackle. He caught a lot of fish with this bait, which I supposed resembled a spinning sprat but was a lot cheaper to use. Another "fun bait" (needless to say, made in the USA) took the form of a small naked female with her hands on her hips, acting as propellors. She spun very well and caught a fish, and is entered in the "Which Bait" section of the game book as a "naked lady". I wonder what they will think in a hundred years' time on reading that? I've witnessed salmon taking a carrot, banana and orange peel, all bearing out the Professor's theory that salmon will indeed take anything when they want to.

Yet another controversial fishing subject is the method employed to remove the fish from the water and onto the bank after being played out. It is really something that is a matter of personal taste and skill, although having said that, there are dangers and risks of losing a fish to all methods. In the spring, when we expect to catch fish 15 lbs or more, I prefer to use a gaff. To my mind it is safer, more positive and the quickest way to get your fish out, especially when fishing alone. I'm not very fond of telescopic gaffs. I've seen too many fish lost when a section pulls right out and the poor fish falls back into the river, breaking the line, then flounders off to die with a piece of the gaff still in its side. My own gaff is an old friend, the head of which was made by the Estate blacksmith many years ago from a round file which he heated, bent to a curve with our traditional three-finger-wide gape, then ground to a needle-sharp point. The metal gaff-head is lashed onto a stout ash pole four feet long, with an old fly line and a cork to protect the point and me!

Next, should you gaff the fish over the top or from below? Walter taught me to go over the top and aim to put the point in midway down by the dorsal fin. In this way the gaff goes into the firm flesh and there is less chance of it tearing out, and anyway his fifty years of experience is a good enough lesson for me. He made the point that in gaffing from below, the gaff is inserted into the soft underbelly of the fish. If it is a big one, the very softness of the flesh renders it liable to rip, and so you lose the prize catch.

One other lesson I have learnt when gaffing a fish is to use a swift, smooth pull and swing the fish well back from the waterside. I have seen, to my horror, fishermen gaff a fish and drop it on the bank only a couple of feet from the edge, saying, "What a lovely...." They don't have time to finish the sentence before the fish gives a couple of flaps and is back in the river with a broken line, lost, to die slowly – quite dreadful. So please, fishermen, take your fish way out in the field, dispatch him with your priest, then, and only then, admire your capture.

There are some people who like to gaff their fish in the head or under the gills. All I ask is please don't ask me to do it. With so much very hard, shell-like protection around the head, I do not like to risk the gaff skidding off the gill-covers and so pulling out the hook. I know gaffing may spoil a portion of the fish, but I would sooner have 20 lbs on the bank and a little of it spoiled than see the fish swimming away. I have an unwritten law that when approaching anyone playing a fish I say nothing unless asked, and I also always ask if the person wants me to gaff his fish, as I can appreciate that a lot of fishermen like to do it themselves. If they do ask me to gaff for them, then I do like *them* to be quiet. There is nothing more off-putting and nervous-making than to have a chap telling the gaffer, "Now. No. In a minute. Don't miss it," and such-like unhelpful orders.

It is tempting fate to say that so far I haven't lost a fish by mis-gaffing, but I do know it as an ever-present hazard. There is one other hazard to gaffing and that usually happens when the fish is too close to the gaffer. Everything is done correctly but the fish, to everyone's amazement, does not come out of the water. The reason is that the gaff touches the fish at the wrong angle and scrapes the side of the fish, impaling one scale on the point. When that occurs you might just as well have left the cork on the point. It doesn't happen very often, thank goodness, but if it does, then check the gaff-head.

I do not like to gaff small fish. The fish, being light in weight, is likely to roll over if enough "jerk" is not applied. If I have to gaff a small fish I like to have the fish a few feet from the bank and keep the gaff low down to the water, in this way putting the point into the side of the fish.

I well remember being called to gaff a fish for Sir Thomas Weldon, a particularly fine fisherman. The water was low and the bank steep, which meant about a four-foot lift to terra firma. The fish was a grilse of around 4 lbs. Sir Thomas played the fish in his usual firm but expert way and finally brought it round, but it came hard under the bank. I could see it quite clearly through the polaroid glasses I was wearing and with no problem slipped the gaff over the fish, lifted it out and bonked it on the head. When I looked at the fish closely, there was no gaff mark to be seen. I must have put the gaff round the fish at exactly its point of balance and lifted it out nestling in the crook of the gaff. I went hot and cold, for I shudder to think what Sir Thomas, gentleman though he was, would have said had I lost his fish. Another lesson learnt!

There is little doubt that a fish can be gaffed sooner than any other method used for removal and I do like to get a fish out as soon as possible, for the longer it is on, the greater the likelihood of the hook-hold giving way. We never know how well a fish is hooked until it is on the bank.

For smaller ones a net is the answer. I use a Gye net with a good deep bag of three-quarter-inch mesh. I've tried small-mesh nets but in fast water they tend to stream out in the current, making the net very difficult to hold and operate. Carrying a large net can create all sorts of troubles. Every thistle or bramble almost seems to have been planted with the sole purpose of catching the net, and the very splendid leather slings provided with the net (at extra cost) are usually awkward to release when needed most and require a degree in engineering to refit after use! I make no claim to have invented this idea for easy net carrying (I may have read of it or seen it somewhere), but it makes for trouble-free handling. The bag of the net is put in a stout plastic cover and secured with a rubber band, which keeps the mesh encased, preventing it from constantly catching in the undergrowth. For the sling I simply use a short elastic "bungee" of the type made for securing luggage on a car roof-rack. They have a large hook at each end which conveniently fits round the landing net, top and bottom. The hooks are very easy to release while playing a fish, and the plastic bag comes off by putting it under your toe, at the same time extending the handle. I have found this manner of net-slinging to be most comfortable and quick in action.

A net is a useful tool to have along with you when fishing early in the season, in case a kelt is hooked and has to be taken from the water, unhooked and returned as quickly and gently as possible, but I don't like cluttering myself up with net and gaff so just carry the gaff and, if I should have a kelt, tail him by hand. This brings me to the "tailer", a device I very seldom use and must admit I do not like. At Broadlands, a lot of the banks are quite steep and require a long lift to get your fish onto the bank. I remember seeing a fisherman playing a big spring fish. He wanted to use his "tailing machine" himself, as I stood by with the gaff in case of accidents. The fish behaved itself and came alongside under our feet, the angler put the wire loop of the tailer over the fish's tail and tightened it. Because of the height of the bank it was necessary to lift the fish about three feet vertically. With the length of the tailer and the fish on the end, the fisherman's arm wasn't long enough to lift the fish clear. The tailer slipped from his hand and the salmon swam off, trailing four feet of tailing machine. Fortunately, the hook-hold was good and the fish was still on, and because of the dragging weight on its rear end, it very quickly came to the side and was gaffed. It weighed 30 lbs.

Tailers do work well, especially on a sloping bank where, once the noose is tightened, one can run back, dragging the fish out of the water. The worst trouble I find with them is that should the noose spring before it is used, it has to be reset. This job really needs two hands and, if a fish is being played and you are on your own, it presents a very tricky problem. I have known this to happen, especially in fast-running water where the force of the current has been enough to spring the loop. There was a very good type of tailer made some years ago that had a positive locking device for the noose and was released by operating a trigger on the handle. This was a much safer method of tailing and did away with the pre-springing problem.

Hand-tailing a fresh fish can be used, and often is. I've done it myself on several occasions but again it's a risky business, especially with grilse whose tails are inclined to fold up and squeeze through the hand like a piece of wet soap. If I have to resort to tailing a fish by hand I usually wrap a handkerchief round my hand first to give a better grip. This method, too, is a difficult one to employ on a river with high banks, which is why we do not often beach fish at Broadlands: we have very few places where it is possible.

By and large, my preference on my own water is to use my gaff, and I've used it on one or two "strong" fish! One afternoon while on my rounds, visiting each beat, I heard a cry for help from an angler who was in trouble. I did a four-minute mile and arrived on the scene to find the fisherman playing a good fish which had run off all his line to the backing. Now the fish was lying doggo well below a group of trees on the bank, spanning twenty yards, which we were unable to get the rod round. I ran below the trees to see if the fish was visible. "He's coming up!" the fisherman cried.

Sure enough, I saw the line going steadily upstream past me, close to the bank, and there was the fish swimming slowly just under the surface. I put the gaff over him and whipped him out. "Got him!" I shouted to the fisherman, who was out of sight.

"Yes, he's still on," the man replied.

I was rather nonplussed, so trotted back to him with the fish, to find him with his rod still bent and obviously still playing one. I eventually gaffed his fish, which was well hooked. The first fish I had taken out had no signs of hook-marks in its mouth, and I can only assume I had gaffed a "follower". Since that day I have often seen a second fish following closely a fish that is being played, as if trying to help the salmon in trouble.

Another time, answering a call for help, I found myself on the opposite bank from the man playing the fish, with the nearest bridge a long distance away. The problem in this case was that the fish had taken the line through a very large bunch of weed and was sulking, with the fisherman unable to do anything but keep a steady strain on him. My advice was to slacken off the pressure, as quite often the fish finds its own way out of the weed if this is done. It didn't happen this time, for the fish just took more line, swimming further away from the fisherman but closer to me. It came up under my bank and I gaffed him out, from the "wrong" side of the river.

These weird and wonderful occurrences make salmon

fishing what it is: an unpredictable sport, fishing for the king of fishes, a great leveller where we all stand equal. It so often happens that a complete tyro can have more fish than the acknowledged expert, as I know from when I first started salmon fishing at Broadlands.

Two very capable fishermen had been fishing Rookery beat all morning and had stopped for lunch when I arrived to see how they were getting on. It was a particularly cold and wet February day and they reported having seen nothing, but as a new-boy learner, I could have a go while they were lunching. I put on a yellow Heddon plug and set off to fish. My first stop was Trees Pool, a delightful piece of water with large plane and beech trees almost touching overhead and forming a great tunnel. I dropped my plug behind the top groyne and let it sink, feeling the vibrations of the bait working transmitted through the rod. The wobbling motion ceased, I struck and a very large fish came to surface and swam leisurely round in a circle, refusing to run or do anything spectacular. Next time he came round close to me I gaffed him, a fine 25-pounder covered in sea-lice. I laid my prize in the grass and went on upstream to try the Cattle Drink Pool. I kept my "lucky" plug on and cast it across the pool. On my third throw, the bait stopped and again I struck, to feel the heavy weight of another good fish. It was an almost exact repeat of my earlier fish. He hardly fought at all, and I was able to bring him to the gaff in under three minutes. That fish too weighed 25 lbs, also covered in sea-lice. I staggered back to the hut with the load of fish to find the fishermen still eating. It was well under an hour since I had left them and there I was, a complete novice with two identical 25-pounders. I can only suppose that both fish had just arrived in the pool, having run straight in from the sea, and were very weary, hence the lack of fight which enabled me to gaff them so quickly. It was a wonderful day for me, but very frustrating for my two fishers.

Chapter Fourteen

WHAT OF THE FUTURE for our salmon fishing, and indeed the salmon? Like all fishermen, I am an optimist. One has to be to go fishing at all! But for the salmon of today, my optimism wanes to a point of near-pessimism. In the good years of the late 'fifties and throughout the 'sixties, there were annual catches of 350 fish and comparatively little fishing effort went in to obtain those results. Compare those years with 1976 onwards, when increased fishing effort was producing an average of 160 fish a season. One cannot help feeling gloomy.

The pressure on the salmon and our rivers has never been so great, or at least we think so. Throughout the salmon-fishing history of the river Test there have been forecasts of doom and despondency, foreseeing the end of salmon running our waters. There were seemingly insurmountable problems at the turn of the last century. Mr Douglas Everett, Broadlands agent, expressed his dismay at the tapping of the chalk springs by the South Hants

Water Company above Romsey in 1902, believing it had permanently reduced the volume of water in the river. He also thought that the continual dredging that went on in Southampton Water, with the churning up that it gets by the passing to and fro of the giant liners, would have a prejudicial effect on the fish and prevent them from making for the head of the estuary as early as they used to. The rate of abstraction now, when translated into millions of gallons per day, would, I'm sure, give Mr Everett apoplexy. It certainly makes my mind boggle. And I wonder what he would think today, for his giant liners of 10,000 tons have been replaced by tankers of 500,000 tons and container ships coming up almost into the mouth of the river, which is still being continuously dredged.

In those days, there was concern that much of the salmon spawn was destroyed by the influx of raw sewage from Romsey. At least that problem no longer exists, but what has replaced it? I think that we are now faced with more insidious things to worry about. I have tried making a complete analysis of our catches since the records began, making graphs beat by beat and also of annual totals, in an endeavour to establish a pattern. The peaks and troughs of the numbers of salmon killed over the years form no readable or forecastable cycle. I have only our own game book to go on which, although it records the catches day by day, cannot record the number of fish in the river.

The problem with endeavouring to form an accurate analysis of salmon runs of past years is that there are too many contingencies to consider and evaluate. Apart from the actual number of salmon in the river, there are so many other elements to be taken into consideration: water height, flow-rate, clarity, temperature. Then there are day-to-day weather conditions, atmospheric pressure, bright or dull days, weed growth or lack of it. All these things we know have a bearing on catching fish. There must be hundreds more factors that we know nothing about. To relate all these causes and effects would, I'm sure, require the use of several computers, even if all the knowledge required were available to feed into them.

So as in the past, and indeed in all things pertaining to

basic fishing, there is nothing new. Are we in fact experiencing a "low" in salmon runs, aggravated by deep-sea netting, on which in a few years' time we may look back and smile as we do now at our forebears' worries? Goodness knows I am not advocating complacency, for I am very concerned at the general apathy, at the talk that "something should be done" and at the almost total lack of action to conserve, let alone increase, the runs of salmon in our rivers. We who fish the rivers of Great Britain must continue to watch and guard our waters against any form of deterioration and should not be afraid to shout loudly to the authorities if anything amiss is found or suspected.

Unless there is a case of sudden pollution, the environmental changes in a river are usually gradual and take place over a number of years, and there is a danger of our accepting lower standards of water quality because the change takes so long. Only half a lifetime ago, there were many good salmon rivers in France, Spain and Germany, now mostly destroyed, and one is hard put to find a river holding any salmon. In France particularly, there has been a great decline in rivers containing fish. Mind you, a lot of this degeneration is due not to pollution but to the fishermen killing almost everything that swims and is catchable.

We took our holidays in France for many years, using our Volkswagen Caravanette to get off the beaten track. On these trips I would usually find a place in the van for a small trout fly rod with a few flies and a net, just in case it should be needed, and strangely enough our camping places were nearly always found near water, either river or lake, which seemed to draw my old "Volksy" like a magnet. During one of our holidays we had left the beautiful camp site at Avignon opposite the famous "Pont" and were travelling east for about twelve miles towards L'Isle-sur-la-Sorgue, when we spotted a signpost marked Fontaine de Vaucleuse. Automatically the van followed the directions. The Fontaine was an enormous hole in the rocky side of a small mountain, and when we climbed up to it, we discovered that it was full of pale green, crystal-clear water, which looked shallow as we could distinctly see the stones on the bottom. We learned that it was in fact over thirty feet deep, and it was at this place around Easter-time that great torrents of water poured from the cavern, emanating from one of the longest subterranean rivers in Europe, fed by the snows of Mont Ventoux.

When we were there, all was tranquillity and the water was only trickling over the sill into a quiet, lovely little river, very similar to the upper Test. Quite naturally we followed the stream down and made camp at a site only a few yards from the water. Once settled, we walked quietly up the river. It really was like the Test. There was a good hatch of fly, the French equivalent of Pale Wateries and Black Gnats, and one or two smallish fish were rising to them. I couldn't resist it, but enquiries to the site manager gave me the information that to obtain permission to fish, I must get the authority from "le maire" who was, I was told, fishing higher up river. I walked upstream to find him, and there he was, in shorts and waders, standing in the middle of the river with a keep tank like a knapsack on his back, using a short spinning rod and casting a very tiny Mepps spoon. He was catching fish – but what fish! None was more than seven inches long, but each one was carefully dropped into his tank, eventually to be part of his supper of fish stew. What amazed and later irritated me was that while he was catching a lot of tiny trout, there were several fish of much better size rising freely in line astern close under the far bank.

The mayor came ashore in response to my call and, having established that I too was a "pêcheur", had a long and rather complex conversation regarding the removal of the quarry. When I enquired of him if he ever used a fly, with a Gallic arm-wave he dismissed this method of catching fish as useless, for all the rising fish were close under the bank and anyway, more fish were takeable on a spoon. I tried to point out that the small trout he was killing would be better left until they had grown to main-course size, but to no avail. His logic was that a fish in the stew was worth two left in the river, which was difficult to argue against. All the time, the good fish were continuing to rise freely about a foot from the far bank. I asked him if he would mind if I tried "la mouche", fished dry, of course.

I used a cast of 1½-lb breaking-strain, tied on a small Black Gnat and dropped the fly above the lower of the rising fish. He took like a lamb, fought like a demon, came to net and weighed just under 1 lb. I took two more and missed a third, the largest fish just over a 1 lb. The mayor was delighted with the fish but I am sure he wasn't convinced that the fly was a good method of catching them. I am equally sure that he would continue to kill every tiny fish he could on his small Mepps spoon. Fortunately I don't think that the mayor's attitude will ever be prevalent in this country.

Many attitudes have changed over the years, both of fishermen and of keepers. The old master–servant relationship which existed in the past is rapidly disappearing and being replaced with a friendliness which can only be good for all concerned.

There has been some great progress over the years in the river-keeping world. In the old days it was 'them and us': salmon-keepers versus trout-keepers and never the twain should meet. Although we all knew each other by name, there was no question of fraternising, let alone helping. In 1970 several of the younger, new keepers got together and we formed the Hampshire River Keepers Association. The idea was *not* to become a union, for all our jobs varied so much in length of water, responsibility, etc., that any thought of a union was out of the question. Anyway, a strike of Hampshire River Keepers would hardly bring the country to its knees! The intentions and aims of our Association are primarily the exchange of the wealth of knowledge available from a group of very experienced men, to further and improve our waters. To quote Izaak Walton, 'Angling may be said to be so like mathematics that it can never be fully learnt.' Of course we have fun too, but the main thing is that now we are all friends and colleagues, ready to help one another in any way possible. For instance, when one of our members was very ill, we formed a working party to cut the weed on his stretch of water, saving him a lot of worry and giving us a great deal of pleasure. That sort of thing just would not have happened in the early days.

I do realise that the price of progress must be paid, but sometimes in my opinion, that price is too high. Many of the large estates that were thriving in our country only a few years ago have through lack of finance been broken up and sold off in small lots for development. It is very sad to see this happen, for many of those estates were part of our heritage, and it grieves me to see it being eroded away to the detriment of the life of the countryside. One can never be sure of the future, but I am happy to think that Broadlands will survive as it, too, moves with the times, although hemmed in by more and more buildings. The Estate forms a keepered island, acting as a buffer between Southampton and Romsey and preventing our lovely little town being engulfed by the crawling city.

For myself, I am content, fortunate indeed in enjoying my work. I most certainly have no regrets about leaving London nearly thirty years ago, except possibly that I should have left the "Smoke" earlier. But then had I done so, I maybe would not appreciate my life in the country as I do. I am privileged to be part of the "Broadlands family" and honoured to have known, to have worked for and with, Lord Mountbatten for twenty-five happy years, and now with Lord and Lady Romsey. I am lucky, too, in meeting so many famous and interesting people in the companionable environment of the riverside fishing world, many of whom have become firm and dear friends.

In my years at Broadlands I have expanded the fishery, some would say making a rod for my own back, but I don't feel that way. Seeing the trout stream develop along with the lake and trout rearing have given me a lot of pleasure as well as hard work. There have been certain sacrifices. Our children were "fishing orphans" for much of their early life, at least on their father's side. The only time I saw them, they were either asleep or in their night-clothes. But I have derived great pleasure from watching them grow up with roses in their cheeks and from being able to give them an insight into nature, seeing animals, birds and fish they would probably never have seen in London.

Through it all, with the ups and downs we have experienced, I am blessed in having Marie, my wife, at my side.

Although a fishing widow for nearly all our married life, spending many hours alone, she has given me constant encouragement and backing without which I would have achieved nothing: in fact we would not even have moved to Romsey.

It has all been worthwhile. My beloved river Test, though changed, continues to flow and long may it do so. I have a deep concern for the future of the river and more so for the salmon that have run it for so many years, but as a final thought tinged with optimism, I will quote Augustus Grimble. A great writer and fisherman, in his book *Salmon Rivers of England and Wales*, written in 1904, Grimble quotes an old friend, a Scottish netsman, who thought that after a period of decline in the numbers of fish caught, "The salmon runs would always make up, ay, and a bit more." Grimble goes on to say of the Test, which he knew well, "Alas! That I should have to proclaim myself a false prophet, for this river is a deteriorating one, and though fish are still to be caught, they are not so plentiful as they were ten years ago."

That was eighty years ago. I'm saying exactly the same thing now, and I pray that I am worrying just as needlessly as Augustus Grimble was in 1904.